# Integrated Circuit Cards,
# Tags and Tokens

# Integrated Circuit Cards, Tags and Tokens

## New Technology and Applications

Edited by

## P. L. Hawkes, D. W. Davies
## and W. L. Price

BSP PROFESSIONAL BOOKS

OXFORD LONDON EDINBURGH

BOSTON MELBOURNE

Copyright © P. L. Hawkes 1990
Chapter 3 © 1990 by The General Electric
  Company plc

First published 1990

British Library
Cataloguing in Publication Data

Integrated circuit cards, tags and tokens.
  1. Smart cards
  I. Hawkes, P. (Peter)
  II. Davies, D.W. (Donald Watts),
  III. Price, W. L.
  004.5'6

ISBN 0−632−01935−2

BSP Professional Books
A division of Blackwell Scientific
  Publications Ltd
Editorial Offices:
Osney Mead, Oxford OX2 0EL
  (Orders: Tel. 0865 240201)
8 John Street, London WC1N 2ES
23 Ainslie Place, Edinburgh EH3 6AJ
3 Cambridge Center, Suite 208, Cambridge
  MA 02142, USA
107 Barry Street, Carlton, Victoria 3053,
  Australia

Set by Setrite Typesetters Limited
Printed and bound in Great Britain by
MacKays of Chatham PLC, Chatham,Kent

# Contents

# Preface

The 'smart' card single chip computer in a plastic credit card shape is widely promoted by its numerous suppliers and their agents as the ultimate microcomputer destined to be carried by everyone everywhere sometime soon.

Why, where, when, questions from prospective card holders amongst the public and the key intermediaries like the bankers, retailers, medical profession, public administrators and telephone companies do not always receive straight answers. The benefits of using smart cards are less tangible than the early costs of introducing systems based on these intriguing devices. In this book we attempt to help the reader resolve the many paradoxes associated with the smart card and its close relatives, the radio tag, the integrated circuit digital memory card, the token and electronic coin.

Amongst the many paradoxes bedevilling the whole subject are the following.

Most of the tens of millions of smart cards now produced annually are not 'smart', more usually they are the humbler relative called the integrated circuit digital memory card. Most of these are used for vending applications like public payphones where an equally cost effective result can apparently be achieved with an optical recording card.

The commonest smart cards produced have on one face of the card electrical inter-connections to the read/write authorisation units. This type of card is the subject of international standards work. However, for many applications these contact smart cards are being challenged by the new contactless radio linked cards such as those available from GEC and AT&T.

But even these new contactless radio linked cards are not as new as they seem. They are predated by the well established radio tag used in the access control field to identify animals, people or goods.

Mars Electronics have shown that it is possible to design an electronic coin having the shape and size of a conventional coin but functioning as a stored value device. There are many other prospective designs of smart 'card' where non-card shapes are preferable for good mechanical and economic reasons. We thus have the paradox that the only real justification for the smart card being card shaped and sized is the transient problem of devising a terminal which will read both magnetic strip and embossed cards as well as smart cards.

Another paradox lies in the claims for smart card security. The card is hailed as the ultimate in security for both access control and as an instrument in financial transactions. In the latter application the smart card is capable of dispensing and recording as data transferred value (equals money). Card stored or emitted files of data, the equivalent of money, obviously require protection from deliberate or accidental misuse both from the authorised card holder breaking the rules and from thieves. To protect card stored data and emitted messages requires data protection measures. These are best based on the applied mathematical techniques of cryptography. The chapter by Dr D. W. Davies describes some of the basics of this most important software area.

Given satisfactory software and economic and durable hardware most application systems based on smart cards remain vulnerable to misuse of a valid card by unauthorised card holders who have stolen or worse still borrowed genuine cards from the authorised holders.

Establishing the cardholder's right to use a given card is currently based on the holder producing the appropriate personal identity number (PIN) or password. Both PINs and passwords can be readily extorted or otherwise obtained from the cardholder's mind or records. Thus although the smart card itself may be secure against many types of misuse limiting use to the authorised holder can be a real problem. Dr J. R. Parks describes the new technology of biometrics which seeks to reduce current dependence on PINs by making measurements on some characteristic of the person such as voice print, fingerprint or handwriting style in order to confirm that he/she is indeed the authorised cardholder.

Some limitations of smart card systems can be overcome by using them in on-line systems where every transaction must be authorised by real-time checks on centrally held lists of stolen and barred cards. The communications infrastructure for a totally on-line system is very expensive. Arlen Lessin's chapter describes one of the new super-smart cards which operate off-line.

For many large scale applications smart cards remain impossibly expensive. To reduce the burden of cost a multifunction smart card has been suggested with a master card issuer franchising space on his card for other card service providers. However, implementing such a system for new payment services such as satellite subscription TV poses substantial administrative and security problems which may delay the commercialisation of such concepts.

In the field of patents smart card ideas have been patented by inventors in a number of countries as well as France. The early use of smart cards will require careful attention to the possible need for licences under some of these patents. Both suppliers and card issuers will need to be meticulous in their study of the published patents and their validity.

Notwithstanding all the above it seems inevitable to the authors that some form of portable personal data carrier will soon come into widespread use in many parts of our society. Whether the smart card as we know it or alternatives such as the optical card, the high density magnetic card or other similar devices will dominate remains to be seen. It is hoped that readers will find answers to some of their questions in this book and that the references given by the authors of the various chapters will lead them to the basic sources of new information on this increasingly important subject area.

P L Hawkes
London
May 1989

# Acronyms

| | |
|---|---|
| AI | Artificial Intelligence |
| ANSI | American National Standards Institute |
| API | Automatic Personal Identification |
| ASCII | American Standard Code for Information Interchange |
| ATM | Automatic Teller Machine |
| | |
| BTG | British Technology Group |
| | |
| CBC | Cipher Block Chaining |
| CFB | Cipher Feedback |
| CMOS | Complementary Metal Oxide Semiconductor |
| | |
| DARPA | Defense Advanced Research Project Agency |
| DES | Data Encryption Standard |
| | |
| ECG | Electrocardiogram |
| EDI | Electronic Data Interchange |
| EFTPOS | Electronic Funds Transfer at the Point of Sale |
| EPROM | Electrically Programmable Read Only Memory |
| | |
| FAR | False Alarm Rate |
| FIPS | Federation of Information Processing Societies |
| FRR | False Rejection Rate |
| | |
| IC | Integrated Circuit |
| ID | Identity; Identification |
| INTAMIC | International Association for the Microchip Card |
| I/O | Input/Output |
| ISO | International Standards Organisation |
| IV | Initialisation Variable |

KB          Kilobytes

LED         Light Emitting Diode
LF          Low Frequency
LMK         Local Master Key
LPC         Linear Predictor Coefficient
LTS         Long-Term Spectra

MAA         Message Authentication Algorithm
MAC         Message Authentication Code

NPL         National Physical Laboratory

OFB         Output Feedback
OSI         Open Systems Interconnection

PAN         Personal Access Number; Personal Account
            Number
PC          Personal Computer
PI          Personal Identification
PIN         Personal Identification Number
POS         Point of Sale
PTT         National Public Communications Authority

Q           Q factor of a circuit
QR          Quadratic Residue

RAM         Random Access Memory
RF          Radio Frequency

RSA         Public Key Cryptoalgorithm (Rivest, Shamir and
            Adleman)

SD          Standard Deviation
SM          Similarity Measure
S/N         Serial Number
SRI         Stanford Research Institute

UV          Ultraviolet
VDU         Visual Display Unit

# List of Trademarks

The following trademarks have been used in the text:

CARL
Cotag
Identikit
Identimat
Innovatron
MagnaCard
Qsign
SIGMA/IRIS
SuperCard
SuperSmart
System 7.5
Talisman
UltiCard
UltraSmart Card
UNO
watermark

Chapter 1

# Introduction to Integrated Circuit Cards, Tags and Tokens for Automatic Identification

### P. L. HAWKES

(British Technology Group)

*In which we discover that the smart card is one of a large family of chip-based artefacts for automatic identification.*

## 1.1  INTRODUCTION

Choosing a title for this book was not easy. People want information on the smart card and its applications. Manufacturers' sales literature is a good starting point but is inevitably biased.

A smart card is commonly understood to be a single chip integrated circuit microcomputer built into a plastic credit card. However most of the smart cards in actual use today are not true microcomputers but nearer memory devices. Many are not single chip, chip cards and some of the best and cheapest of these are not even card shaped!

In fact the smart card is but one of many integrated circuit-based data carriers used in a wide variety of computer systems to help identify people, animals, plants, things, messages, events and places. Indeed it is easier to define what is not a chip-based portable data carrier than to produce an overall definition. Concentrating on automatic identification seems to the author as good a basis as any.

Another surprise is that the history of automatic identification via a personal portable data carrier based upon a digital integrated circuit device goes back to 1968 or earlier. The various designs now available reflect the different origins of the data carriers concerned and their prime applications – anti-shoplifting tags, magnetic stripe identity cards, vending cards, pocket calculators etc.

The achievement of M. Moreno and his French licensees and partners has been to focus worldwide commercial attention of one particular class of integrated circuit memory cards. This is the class of miniature artefacts shaped like a standard plastic credit card, having the same dimensions and containing hardwired or programmed logic as well as digital storage, i.e. the so-called 'smart' or 'intelligent' memory card. In the early 1980s Roy Bright introduced the adjective 'smart' to describe succinctly the essential characteristics of the single chip microcomputer card. His more recent definition distinguishes between the 'active' smart card and 'passive' smart cards. The important features of the former are described in Chapter 2.

In this initial chapter, I will attempt to survey all the silicon chip-based technologies and the perceived needs propelling their creation and uses.

## 1.2   BASIC FORM AND FUNCTION

Integrated circuit cards, tags and tokens are components in distributed computer and telecommunications systems. Basically they exploit the low cost high density digital storage capacity of integrated circuit memory chips usually, although not invariably, in association with control circuitry known as logic.

As our children are probably now taught in school, integrated electronic circuits are more or less complex arrays of transistors, diodes and other circuit elements and their wiring interconnections formed by printing, diffusion and other processes within a single die or chip of silicon or other semiconducting crystal.

By selective contact printing and etching device, structures down to a few ten millionths of an inch wide are created and enable the resulting chip to record information and process it very rapidly.

With rapid and continuing progress since the early 1970s, integrated circuit making has progressed until today, a single chip IC some half inch square by a few thousandths of an inch thick, can record up to several million bits of digital data as an electronic charge pattern. The microcomputer's logic equivalent can process data at 20 million or more operations a second.

Further increases in information recording density and data processing speed are expected. Made in arrays on six inch diameter wafers, the chip itself sells for a dollar or two.

Like its competitors, magnetic discs and cards and optical discs and cards, the IC chip presents the technologist with a new information recording medium. Using low cost integrated circuit memory as the basic medium, the system designer has a new tool or instrument with which to disseminate and record information *in a system*.

The basic functions enabled by the IC memory chip are the storage of a 100,000 or more bytes (characters) of text or data and their emission or recording in less than a second. Unlike the optical and magnetic media, on-chip logic permits memory access to be controlled autonomously from within the chip. The implications of this are far reaching as will be described below.

## 1.3 GENERIC APPLICATIONS

At the present state-of-the-art, the basic form and functions of various IC cards, tags and tokens can conveniently be classified as shown in Table 1.1. The exact form of memory used in these devices varies widely from UV or electrically reprogrammable memory devices to battery backed RAM (random access memory). Particular products and designs categorised in Table 1.1 are best suited to specific applications. These are summarised in Table 1.2.

**Table 1.1** Integrated circuit cards, tags and tokens

| Type | Typical capacity (bits) | System interface (s) | End-user/card holder interface |
|------|------------------------|----------------------|-------------------------------|
| Radio tag | 64 | RF coupling | Via system interface |
| Memory only card | 16K−1M | 6−8 electrical contacts | Via system interface |
| Wired logic 'smart' card | 256 up | 6−8 electrical contacts | Via system interface |
| Programmable logic 'smart' card | 8K up | 6−8 electrical contacts | Via system interface |
| RF programmable logic 'smart' card | 8K up | RF coupling | Via system interface |
| Active smart card | | | |
| (a) Smart Card International 'UltiCard' | 8K up | Direct by contacts or indirect by card user | Direct by onboard display and keyboard |
| (b) Visa 'Supercard' | 8K up | Direct by contacts or indirect by card user | Direct by onboard display and keyboard |
| (c) NPL 'Talisman' token for RSA messages | 30K up | Direct by contacts or indirect by card user | Direct by onboard display and keyboard |

**Table 1.2** Typical applications of integrated circuit cards, tags and tokens

| Type | Actual or proposed application |
| --- | --- |
| Radio tag | Identification of specific people, animals, places or goods |
| Memory only card | Distribution medium for computer programs and data |
| Wired logic 'smart' card | Vending card for making calls from public telephones, etc. |
| Programmable logic 'smart' card | General purpose including credit and debit card for use in on line and off line payment systems and 'electronic wallet' |
| RF programmable logic 'smart' card | As above |
| 'Active' smart card | (a) off line payment systems<br>(b) patient data cards in medicine<br>(c) signing and encryption of electronic mail documents<br>(d) metering of the use of gas, water, electricity, TV, public transport etc.<br>(e) logging of events e.g. accesses to premises |

## 1.4  SYSTEMS

The smart card, tag or token is an instrument, usually the 'key' instrument in a complete system designed to provide a service to the end user, i.e. the person carrying the instrument.

The service provider operates and sometimes designs the system. The appropriateness of the particular card, tag or token for a particular service is measured in terms of speed and ease of use, security and cost. Cost reflects both purchase price and cost of use.

Systems are classifiable into two main types − public and private (see Table 1.3). Private systems are intended for use by a closed user group, typically the employees of the organisation operating the system. An access control system for a company's premises is a common example.

Public systems are designed for use by members of the general public, qualified only by a virtue of being customers of a particular bank or users of a particular public service such as the payphone system.

The important public systems are those like credit cards and charge cards which operate internationally as well as nationally. The relevant

**Table 1.3** Public and private IC card, tag and token systems

| Class | Card population | Card/terminal ratio | Role of standards | Terminal security and price |
|---|---|---|---|---|
| Private system | tens to thousands | low (10:1 up) | Useful | Both high |
| Public system | millions | high (50:1 up) | Quintessential | Both generally low |

standards are therefore evolving from suppliers' and service providers' standards into international ones via the appropriate national standards bodies, INTAMIC and similar bodies.

Cards, tags and tokens appropriate for public systems tend to be ultra simple to allow customer activation. Low cost is also essential and generally possible because of the large number of standard units involved. This makes them attractive candidates for use in those private systems where the functional limitations can be tolerated.

Operating generally on a single site, over a restricted geographical area or via private networks, private systems can usually afford to have on line real-time telecomunications with each card terminal in constant touch with the system's control centre. This makes the management of card security relatively easy compared with public systems. However, some 'open' sites like hospitals and hotels present particular difficulties associated with the ever changing authorised user population and the risk of attack by criminals and vandals.

Public systems for payment (revenue collection) and the disbursement of money (revenue distribution) are obviously subject to misuse both by legitimate card holders and imposters. This makes on line real-time notification of lost or stolen cards and of account abuse highly desirable. Quick circulation nationally or internationally of 'hot card' lists is however expensive so most systems incorporate a degree of off line operation. This is also of course vital to allow the authorised card holder to obtain some element of usage even if there is a telecommunications failure. Just imagine a bank which told its current account holders they could not use their cheque books because the bank's computer network had problems!

Terminal security and cost are big issues in both types of system. Many of today's terminals are in well protected environments e.g. ATMs on bank premises. Their operation by customer activation can therefore be trusted, This will not be true of many retail shop terminals. Recent scares about computer program 'viruses' demonstrate widespread concern in the industry about the difficulty of trusting personal computer-based terminals.

This may cause a re-evaluation of the security needs and precautions taken when designing, installing and operating PC-based card systems.

A good solution may appear with the new 'active' or super-smart cards (Table 1.1). Having their own keyboard and display this class of device need not rely on a trusted terminal for most of its operations.

## 1.5  SOFTWARE AND PROTOCOLS

Software includes the programs governing the operation of a programmable electronic device such as the 8-bit single chip microcomputer in a typical 'conventional' smart card. Also included is the operational data which 'personalises' a card, tag or token to the individual authorised end user and the service providing organisation. This data may be programmed into the various types of memory mentioned above, expressed as a wiring pattern (masked programmed) or via fusible electrical links.

Protocols are essentially the rules of conduct by which the card, tag or token communicates with its system or other similar devices. They can be designed in as hardware or software.

Much of the available on-chip memory can be consumed by a stored program for control of the operation of a programmable device. Thus for any very large scale application a bespoke, hardwired solution consumes less chip area and is therefore cheaper. The pay telephone card is a prime example.

## 1.6  SECURITY THREATS AND THEIR CONTAINMENT

Since the basic purpose of an IC card, tag or token is to identify the bearer to a system, security lies at the heart of all applications. It is therefore not surprising that improved security against misuse by card holders, authorised as well as unauthorised, is often the main selling point for these components. This emphasis has reached the point where the smart card for example is sometimes presented as a panacea for all manner of retail banking and access control systems.

A project sponsored by the author's employers and carried out by the Data Security Team at the National Physical Laboratory, Teddington, has examined the security of smart cards and systems, identified threats from the likely sources and devised appropriate new hardware and software technology to contain the dangers. A prototype version of NPL's 'Talisman' device was developed with the help of Texas Instruments Ltd. Full details are given in Chapter 6. It is described as an integrated circuit 'token' rather than a super-smart card because the recommended size is greater than a credit card and the shape can differ to suit the application.

The main points relating to smart cards used by people are as follows. The card is essentially used to support the card bearer's identity claim. Once read in an authorisation unit (terminal) and accepted as valid the system allows the card bearer to complete a requested transaction. The relevant transactions include:

- Purchase of goods or services
- Access to private premises or computer resources and data
- Sending or receiving telecommunicated messages of value

The threats come from misuse by the authorised card holder, misuse by an unauthorised card holder or where there is collusion between such parties.

Abuse cannot be entirely stopped except at uneconomic cost so a well designed smart card application must contain it. This can be done for example by denying future services to an authorised card holder who has abused his privileges or by catching a thief either in the transaction or later via an audit trail.

The main basic security weakness of the conventional smart card is that it can be stolen and used by an unauthorised card holder.

The established way to guard against this is to only allow card activated transactions where these are supported by the card holder producing a valid PIN (Personal Identity Number). However this PIN must be entered via the keyboard of an authorisation terminal. As already stated this terminal may not always be trustable. If it is bugged a criminal can discover the secret PIN without the card holder's knowledge, copy or steal his smart card and then obtain access to money, goods, services etc. from his account with the card issuing organisation.

NPL's solution to this with its 'Talisman' IC token is to provide a keyboard on the token itself. With a trusted display on the token this keyboard makes the token's use less vulnerable to untrustworthy terminals. Similar solutions are being pursued by Visa and Smart Card International (see Table 1.1. above) under the terminology 'active' smart card.

For many applications of smart cards and tokens, messages need to be sent from the card to a remote mainframe over an insecure network. To prevent eavesdroppers abstracting, delaying, altering or inserting messages the technique of cryptography needs to be employed. Chapter 8 describes these.

The Talisman token incorporates encryption means for generating a cryptographic version of messages sent from the token to remote computers or other tokens such that the message cannot be read by any but the intended recipient and he can authenticate that the message must have come from that token and no other.

PIN details and other confidential data stored in a smart card, passive or active, or in an IC token can be discovered or altered by unauthorised investigation of the IC memory and its data contents. Data alteration is especially likely for smart cards and tokens used as 'electronic wallets', 'cheque books' or meters. Attacks can be logical (via the contacts etc.), electrical (in the same way or by radiation detection) or physical by opening up the unit and reading the data stored therein. Tamper proofing is possible but very costly so most commercial products are best described as 'tamper resistant'. Known means include sensitive 'triggers' which wipe out card stored data when tamper attacks are detected. Easily broken wires buried in a resin potted chip module are one example of triggers. These can be rendered ineffective by deep freezing so they are not a panacea.

Another area of vulnerability is the PIN itself which can be guessed as well as stolen. This has led NPL and others to investigate the uses of so-called 'biometric' techniques whereby some measurement is made of a personal trait of the authorised card holder and compared with an authenticated card stored reference.

The operation of a biometric device is analogous to the 'eyeball' comparison of a handwritten master signature on for example, a conventional credit card with a new specimen produced on demand for a bank cashier or shop assistant. Not surprisingly then automatic signature verification has received a good deal of attention from NPL, SRI/Visa, De La Rue, Thomson and others. It is a well accepted and legally binding commitment to a transaction. All these designs exploit handwriting timing and rhythm as well as signature outline. Such invisible 'dynamic' signature characteristics are very difficult for a forger to reproduce and quite easy for a computer to analyse given an accurate handwriting encoder.

Chapter 7 describes the current state-of-the-art in biometrics including signature dynamics, hand geometry, fingerprints, retinal and hand blood vessel scanning and speaker verification. To be used effectively with a smart card or token the biometric validity decision must be made by the on board microcomputer using locally stored reference data.

Promising solutions leading perhaps to a biometric smart card are being worked on by a partnership between NPL, the British Technology Group and several equipment suppliers and card issuers. These solutions may soon result in a cost-effective biometric smart card or token. Meanwhile an interesting compromise is to store 'mug shots' in digitised form, in a smart card. Human operators of manual terminals can then compare the card stored 'mug shot' with the claimant's appearance and then authorise or deny the requested transaction. This should prove a useful compromise for some markets like physical access control. Clearly it is inappropriate for markets like self-service banking and shopping.

**Table 1.4** Choosing IC card media for access control

| Application | Example | Key feature | | | | |
|---|---|---|---|---|---|---|
| | | Smart Card | Super-Smart Card (Token) | Memory IC Card | Radio Tag | COMM. NIS |
| (A) Logical access | (1) Reading, writing or erasing financial & medical data in databases | Program controlled data store gives versatility | As smart card but more secure | EPROM versions give audit trail | Hands free (passive) operation | Tag also usable for physical access (B) below |
| | (2) Electronic Data Interchange including EFTPOS | Message integrity by private key cryptographic s/w | Message integrity by public key cryptographic s/w | — | — | Latent demand for EDI should generate a substantial token market |
| | (3) Point to multipoint data distribution | SDI by private key cryptographic s/w | SDI by public key cryptographic s/w | — | — | Teletex & satellite opportunities |
| (B) Physical access to premises & sites | (1) Employee access | Data store holds personal & biometric details | Secure PIN validation | Store big enough for mugshots | Hands free (passive) operation | — |
| | (2) Employee location by zone | — | — | — | Passive operation | Tag identification by terminals at zone boundaries |

**Table 1.4** Choosing IC card media for access control

| Application | Example | Key feature | | | | |
|---|---|---|---|---|---|---|
| | | Smart Card | Super-Smart Card (Token) | Memory IC Card | Radio Tag | COMM. NIS |
| (C) Physical access to routes (travel) | Road tolls, airline & train season tickets | – | – | EFROM based security | Passive operation | Moving vehicle problem for tags |
| (D) Telecommunications | Public telephone cards | Rechargeable | Rechargeable | EPROM based security | – | The leading application in the world (20–30 million) |
| A-Z Multifunction cards or tags, | Employee access, work log & discount card | Program controlled datastore gives versatility | Power and security for most purposes | – | – | Super smart card/ RF tag combination would meet all listed needs but so would suitably programmed CT2/ pager device |

## 1.7  OTHER DEVELOPMENTS

Before the ISO standard smart cards are established internationally new designs are appearing with alternative or additional features to open up new applications.

Chapter 3 describes the GEC ic card with its secure low cost RF coupling method for card to terminal interaction.

Two other developments worthy of note come from the opposite ends of the product spectrum of Table 1.1.

The humble radio tag has now fully established itself as a viable solution to the access control problem (Table 1.4). There are over fifty suppliers worldwide. In this country Cotag and its competitors have delivered hundreds of systems to the smaller organisations with a need to restrict site entry to a few hundred employees and some authorised visitors. The systems work well and are cost-effective. John Falk of Cotag describes radio tags and their manifold uses in Chapter 4.

## 1.8  FUTURE PROSPECTS

As the still fledging industry matures there seem to be two opposing tendencies. The first is to migrate towards very low cost standard devices manufactured on a huge scale.

At the opposite end of the spectrum are the active devices like the NPL's Talisman Token. In the author's view these different approaches will coexist.

There may also be scope for the integration of the identification and metering functions of the active smart cards and tokens to be integrated as software into other products like conventional and portable terminals and telephones.

Chapter 2

# Smart Card Technology – A US Pioneer's Viewpoint

ARLEN RICHARD LESSIN

(Chairman & President, Lessin Technology Group, Inc)

*The early pioneers were visionary, seeding the not yet existing market for a then unknown technology. Those activities, however, are now making possible diversified and economically feasible applications.*

## 2.1 INTRODUCTION

The smart card entered the US very quietly in 1980. Drama followed quickly, but this story will have to be part of another book. The event occurred at a November international meeting called INTELCOM '80. The place was the Los Angeles Convention Center. Amidst chromium tubes in angular designs and free-flowing vintage champagne, the French government introduced its new telecommunications capabilities and some other technology developments. The main exhibit area was generally quite impressive.

The smart card (Carte à Mémoire) display was not impressive – composed of a few mock-up 4KB capability cards and a non-working prototype terminal sitting casually on a red table cover. This exhibit was less than centrally located. Having talked with some of the attendees later, none of those who passed by or stopped at that table saw the technology displayed there and believing they had seen anything important. This writer was apparently the major exception. Having visited the exhibit and being immediately captured by its possibilities, a light switched on for me. That conference marked the start of my long involvement in the technology, including two years as special consultant to the French government, introducing smart cards to the US. Whether I have been justified in my faith is still in the proving stage, however, in 1989 the future appears bright – but not yet here. Therefore I have decided to meet the issue head-on by establishing the Lessin Technology Group, Inc. (TLC). This New York based consultative and systems organisation was mandated to accelerate the acceptance of smart card and related technologies in the US and internationally.

## 2.2  EARLY DEVELOPMENT

In 1980 the US government had already granted the five key US patents on smart card technology to Innovatron. That company was founded by French financial/technology journalist, Roland Moreno in 1974 after he invented the key elements of the smart card. At the time, international promotion of the technology was in the hands of Intelmatique, the French government's international technology group, which was responsible for promoting the introduction of the smart card along with several other French products. These included videotex systems, electronic directory systems, low cost consumer terminals, low cost facsimile terminals, telewriting, and audiographics teleconferencing. The key elements, however, proved to be videotex and smart cards; and at the time, videotex was clearly the Government's central concern and smart cards a quite secondary one. In 1980, anyone predicting that nine years later the smart card would be more prominent than videotex was considered either a visionary or, more likely, less than observant of apparent realities.

In France, the banks gave the production version 8 Kbit EPROM smart card early heavy support. To this day, as a result, its potential financial applications are the primary emphasis in much of the literature. But from the start, those involved in the technology saw that it had potential uses that went well beyond the financial industry. For example, it could carry personal medical history, give personal access to confidential information, or serve as a key carrying one's retinal scan or digitised fingerprint or photograph to provide access to areas of great physical security. Philips Data Systems, one of the original three French card and terminal manufacturers licensed by Innovatron, published a descriptive graph in 1982 suggesting that smart cards could be used for payment cards, prepaid cards, subscriber cards for telephones and transport cable TV, authorisation and access, guarantees and service for automobiles, applications and social security services. Some of those applications – and a few not on the Philips list – have been implemented recently. So the ideas behind them aren't all new; it is rather more a signal example of use beginning to catch up with vision. Vision is always parent to the reality.

It took nearly two years for the smart card to formally surface again in the US, although there had been substantial media attention to its potentialities and much work behind the scenes on its behalf had been done. The year 1982 is significant in smart card history for several reasons. In France it was the year that the first home banking system using smart cards went into operation, its first terminal installed in an apartment in Velizy, a Paris suburb, and also the year of the first major Point of Sale (POS) trials. Two trials also started that year in the US.

- First Bank Systems of Minneapolis ran a small smart card trial involving 10 North Dakota farmers, in conjunction with a home banking system videotex trial.

- The US Department of Defense issued a few hundred cards to soldiers at Fort Lee, Virginia. The cards were to be used for identification at post entry and exit points and at the post exchange (PX) where US soldiers and their families can buy goods at greatly reduced prices. At the time, the US Army was looking for an alternative to the cardboard photo ID cards it had been using for generations. The smart card was one of several technologies tried out at different military bases in the US. The smart cards were used in special point-of-sale terminals. These first US experiments did not lead to substantial distribution of smart cards. The Department of Defense in effect made no decision on the smart card – the program started at Fort Lee was shelved. However, currently the Pentagon has been considering issuing a purchase order for several hundred thousand to millions of cards. However, this interest has not yet been implemented.

These were the only visible US trials of smart card technology. In general, US banks were resistant to the technology, essentially because of their major commitment to magnetic stripe implementation. However, several other important activities were actually going on behind the scenes. Chase Manhattan Bank, Security Pacific Bank, American Express, Bank of America, and the US Department of Agriculture, among others, launched major research projects into the potential uses of smart card technology. Some of those experiments have led to new applications of the cards, while other organisations are still waiting and watching the technology develop (including experimental laboratory development and limited test implementations of a non-contact version of the technology by AT&T among others). Constraints to acceptance in the US included limited memory, requirement for readers/terminals at all transaction locations, high cost, no reusability, and the prejudice against foreign technology – the NIH or 'not invented here' syndrome.

Throughout this time, most of the technological development of smart cards was going on in France and, to a lesser extent, Japan. US companies were basically playing with the French technology and often trying out applications that were already in commercial development in France. All of these French and Japanese cards were 'passive' cards, based on a new terminal infrastructure being implemented. Some initial reasons for US slowness to implement smart card technology have been non-updateability of most cards, commitment to magnetic stripe card systems, the need to supply power to the card via an originally costly smart card terminal/reader infrastructure, and the fact that the applications software was *not* in the card, but in the terminal or a PC or computer beyond the terminal. This made changes costly and difficult. That requirement changed in October 1985 when Visa announced its program to test the Super Card – a value added, multiple function, large memory, charge card of the future. It was to be an advanced smart card with its own built-in terminal – including a keyboard, display screen and battery power all in a credit card package. The applications software resides inside the card in

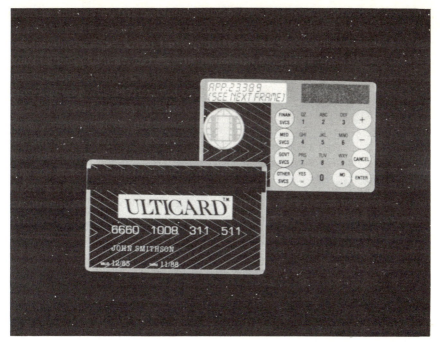

**Fig. 2.1** The UltraSmart Card (formely Ulticard).

the SmartCard International (SCI) UltraSmart type of Super Card and in its SCI MagnaCard (conventional type). It was July 1985 when SCI proposed the A. R. Lessin concept for such a card to Visa. In November this new smart card had 64 Kbits of reprogrammable memory, a two-line readable display, alpha-numeric keypad, and battery power. SCI and Visa signed a development contract in January 1986 and SCI delivered the first working prototype Super Cards at the Visa International board meeting in San Juan, Puerto Rico in May 1986, only 16 weeks after the contract was signed. SCI calls this card, UltraSmart Card, (shown in Figure 2.1), Super Card being one version and an increasingly generic name for this version of smart card technology (although 'SuperSmart' has been registered by Visa). Since then, SCI has worked with Arthur D. Little, Texas Instruments, and other American companies in continuing to develop the UltraSmart Card technology and to find new applications for it.

## 2.3  NEW GENERATION SMART CARDS

This new generation of self-contained smart cards is a major advance in the technology. It radically reduces the need for terminals -- although it can be compatible with most imprint, magnetic card swipe, smart card, and other point-of-sale terminals. (The present generation of this kind of cards are in the process of becoming compatible with ATMs.) To make

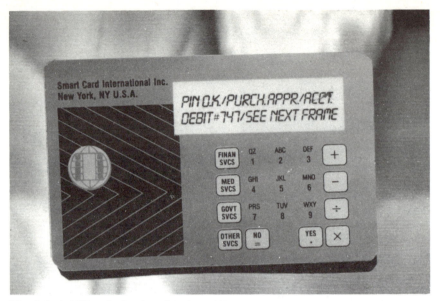

**Fig. 2.2** A PIN unlocks the UltraSmart Card for purchase use or access to other
functions, services and multiple applications.

a purchase, the user unlocks the card with the proper personal identification number (PIN Figure 2.2). The clerk enters the relevant information
on the card's keypad regarding the product being charged. The processor
checks the credit balance and displays an authorisation code which the
clerk copies onto a standard credit slip, or enters into an electronic terminal, for transmission to a data bank.

UltraSmart Card is much more than a simple smart card with a built-in
screen. It is a complete portable computer − informally named 'The
Pocket PC' − essentially credit card size. Previous versions of the smart
card have suffered because they had no way to address memory logically
in the way it is done in standard computers. Instead all data has to be
stored and retrieved by physical address. Thus users could not simply tell
the card that they wanted to see such and such a data file, as they can on
a computer. Instead, the card issuer − or whoever designs the retrieval
software − must have intimate knowledge of the arrangement of the
card's memory and must tell the card the precise physical addresses to go
to get the data. This has made creating application software much more
cumbersome and expensive.

UltraSmart Card − and SCI's conventional smart card, MagnaCard −
were the first smart cards to implement a full operating system, although
recently there have been others. However, the proprietary SCI system,
CARL (CARd Language), handles all memory management in the same
way the operating system of a normal computer does. Thus data can be
accessed by logical memory management, in the same way the operating

system of a normal computer does. Data can be accessed by a logical file name, regardless of where it is stored in the computer. This greatly facilitates development of application software. It allows several different applications – such as in the Visa version to act as a credit card account, an electronic checking (debit) account, and an appointment calendar or address file – to share a single card since the data for each application can be protected from the others. The UltraSmart Card version extends this multifunctionality to medical, government, insurance, educational, professional and virtually an unlimited variety of other applications. In fact its developers anticipate that this type of technology will normally be used in this manner. Also if desired, two or more applications can share a single data file even further expanding capabilities.

## 2.4 FINANCIAL USES

MasterCard in 1984 had already announced its own tests of the traditional smart card technology. In late 1985 it mounted tests of both French and Japanese versions of the 'passive' technology involving thousands of cards in point of sale (POS) applications in Columbia, Maryland, and Palm Beach, Florida. One of MasterCard's main reasons for testing the smart card was its perception of existing and projected credit card fraud. An example of card fraud would occur when someone steals your card along with your wallet in New York. He then flies it to California where the stolen card is used illegally for three days to run up hundreds of dollars in bills for items which are then sold on the black market. This has been a growing problem and is potentially a big business, possibly involving organised crime in the US. MasterCard blames it in part for $900 million in Visa and MasterCard losses in 1984 and has estimated that those losses will exceed $2 billion in 1990 – more than the national debt of some countries. MasterCard was attracted to the smart card because of its superior security, which it believed would stop most of the fraud.

Unfortunately, the MasterCard tests – which were intended to determine whether consumers would accept the technology – proved inconclusive. This was not because the cards or terminals failed or because consumers didn't like them. It was because those running the program did not adequately educate either card holders or retailers in using the technology properly and also the cards and terminals were not adequately integrated into the POS system. This is, I believe, an absolute requirement in regard to utilising smart card hardware and software. They must be integral parts of the overall user system. This caused MasterCard to pull back. In fact both MasterCard and Visa reached the conclusion that they must cooperate in the introduction of any new charge card technology. Installing millions of point-of-sale terminals and training millions of store clerks to use the new technology is too expensive a proposition for either to do alone.

Unfortunately the two companies disagreed − and continue to dis-
agree − on the basic reasons for being interested in the technology.
MasterCard remains convinced that the elimination of credit card fraud
and reduction in credit losses will be enough to justify the expense of
converting from magnetic stripe to smart card technology. Visa sees
combating these problems as a secondary issue. It is mainly interested in
providing cardholders with new services, and it is not willing to go ahead
in a major implementation of smart card technology unless it is convinced
that these extra services will generate the revenues to pay for conversion.
Therefore, Visa's Super Card project concentrates on adding new services
such as electronic checking, conversion of currency into international
denominations, electronic note pad, calculator and clock. Also, elimination
of the smart card terminal infrastructure is very important for Visa,
because it changes the worldwide economic justification for the payment
function drastically. The objective is generating profits, **not** new bank
technology.

The real answer, of course, is that both organisations are right. US
banks are facing a business crisis caused by the deregulation of their
industry and the entry of dynamic new competitors who have already
taken away a large amount of their traditional business. They need to
offer new services in order to hold onto the business they have left. At
the same time credit card fraud is reaching material proportions. If it isn't
stopped, it will become a major problem for the US banking industry in
the 1990s. The UltraSmart 'active' card could provide solutions to both.

A 1987−88 smart card study performed by Booz-Allen-Hamilton, the
large US consulting firm, was funded jointly by Visa and MasterCard.
Focusing on the constraints and costs of implementing conventional smart
card systems, it put a damper on short-term ultilisation of this form of the
technology. However, it did not address the facts of UltiCard technology
and the opportunities it offers in requiring only limited reader infrastructure (being a reader as well as a processor and memory device itself) and
therefore radically reducing widespread implementation costs. A current
Frost and Sullivan study predicts a $20−25 billion business in the mid
1990s.

Unfortunately, in late 1989 the Visa/MasterCard debate shows no signs
of resolution, and blocks full-scale inplementation of either standard
smart card or unified (UltraSmart Card) technology in the US financial
industry. The large banks and other major credit card companies such as
American Express and Carte Blanche adopted a wait-and-see attitude,
also probably put off in part by the initial cost of converting from magnetic
stripe technology. What will change their positions will probably be recognition of such factors as the economies of reuseability and longevity of
smart cards, off line Super Card (UltraSmart Card) capabilities, the cost
of credit card fraud, and the cost of the phone lines needed for on line
credit authorisation that is the main present day (1989) answer to fraud

growth, – led by identified opportunities for new revenue sources. Thereby, the pressure to adopt the alternative, vastly more secure, totally off line system provided by smart card technology will likely increase until it portends to become irresistible even to the most conservative and change-resistant bankers and banks.

## 2.5 AGRICULTURAL USES

While development of strictly financial uses has slowed, other US applications have actively developed. For instance, in the three years (1986–1988) the US Department of Agriculture (DoA) has instituted a program using smart cards to track production quotas for peanut farmers. Under US law, farmers are given yearly quotas on the amount of peanuts they can sell. If they exceed those quotas, they are heavily fined. Because farmers typically do not bring all their crop to market at once, or sell it all to one buyer, it is difficult for farmers to make sure of what they are permitted without exceeding their allowances. Copies of the records of each sale used to be mailed to the DoA where they were reviewed and an updated statement of the remaining quota mailed back – a process that often took two weeks – during which time the farmer may well have made further sales. Thus farmers could easily make costly mistakes.

In 1985–86, the Department of Agriculture began issuing conventional smart cards to farmers in a pilot program that has since grown into the first major full-blown implementation of the technology in the US with 58,000 cards issued to farmers. Each card contains the quota for the farmer to which it is issued. Each time farmers sell peanuts they deduct the amount they sell from the card using standard point-of-sale terminals. This allows everyone involved to know whether the sale falls within the allowable quota and how much the farmer has left on his quota immediately and accurately. The quota program, which is designed to benefit the farmers, has been made much more efficient and most quota overruns eliminated. This program is now being supplemented by a similar one that will apply to tobacco farmers.

## 2.6 SECURITY USES

Conventional smart cards are also finding a place as electronic 'keys' to high security areas such as mainframe computer installations of such companies as Bank of America and Royal Bank of Canada, among others. Smart cards are excellent for this because they can carry multiple security levels. For instance, the SCI MagnaCard can contain up to 22 different keys or levels of security, each of up to 254 characters. The National Security Agency (NSA) will utilise these capabilities as part of a security access program scheduled to involve one + million cards from three US suppliers in the 1989–92 time frame. These can be nested so that they would have to be entered in a specific order. Also, the card can

be programmed to refuse to operate after a predetermined number of incorrect consecutive tries are made at entering a key (usually three). Furthermore, smart cards can carry biometric information such as digitalised
photographs, fingerprint or retinal scan data for use with very high security
devices. The smart card is capable of holding this information in a very
secure manner, making it extremely difficult to alter or forge.

**Fig. 2.3** The SCI MagnaCard. This example is of a traditional type of smart
card, issued in other forms by various suppliers. It can contain different sizes of
fixed or eraseable memories. It requires a reader/terminal for access.

## 2.7  MEDICAL USES

UltraSmart Card type technology is also finding new applications, particularly in medicine. Modern health care may be the most specialised,
decentralised, and information-intensive of all human activities. Better
information at the right time and place means better health care. More
efficient information technology means lower costs.

Medical and health care professionals have been quick to see the
possibilities of smart cards, to improve care and contain costs. UltraSmart
Card, for example, has the memory capacity to store and revise extensive,
dynamic, individual medical records. UltraSmart Card provides complete
multilevel security, allowing open access to emergency information while
preserving individual privacy and protecting doctor-patient confidentiality.
UltraSmart Card's stand-alone capability is ideal for self-monitoring
regimens, home care, regular medication or treatment regimens and
other outpatient programs and services.

The Methodist Hospital's Institute for Preventive Medicine and the
Baylor University College of Medicine, both in Houston, perhaps the
world's largest medical complex, developed and tested a medical systems
UltraSmart Card in partnership with SCI (Figure 2.4). This multifunction
card was designed to be used by patients with heart disease, diabetes and
hypertension as part of a medical regimen. Selected over-weight patients
could also use it as part of a self-monitoring regimen designed to modify

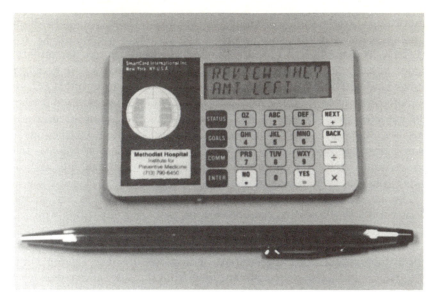

**Fig. 2.4** Medical systems UltraSmart Card.

behaviour patterns associated with overeating. This program could be further enhanced by a US government National Institutes of Health Research study grant to SCI. Patients could use the card to keep track of food intake, exercise, weight, medication, use of alcohol or tobacco and behavioural correlates (stress, boredom) between clinical visits.

Dr J. Alan Herd, formerly Medical Director of the Institute for Preventive Medicine, described the potentially significant impact of 'computer-in-a-pocket' technology on the success of health-related behavioural modification. During the past 15 years, health professionals have developed programs to help control many chronic diseases. In addition to scheduled medication, most treatment programs require patients to keep written records of health-related behaviours. Referring to these programs, Dr Herd praised the increased reliability and accuracy of UltraSmart Card type records and the capability to offload those through a PC interface for graphic presentation. 'The immediacy of feedback and the precision of reports greatly enhance the opportunity for educating and improving performances of subjects when changing dietary patterns in an important part of the medical regimen,' he said.

InfoMed, the largest US company committed to providing custom information management systems to home care agencies nationwide, seeks to reduce health care costs and free maximum staff for patient care by increasing administrative efficiency. Current InfoMed services include distributed data processing systems, advanced home care information systems, stand-alone business application software, personal computer systems, and a handheld nursing computer – potentially one of the first important UltraSmart Card type of applications. There are 10,000 home nursing agencies in the US.

The UltraSmart Card type of Nursing Computer could perform two major tasks for each nurse — daily scheduling and patient care programming. Updated daily via a read-write interface with a local office PC, the Nursing Computer would carry a complete plan of the prescribed medical regimen for each patient to be visited that day including medications, dosages, frequencies, possible side effects, and handling instructions. It would also contain patients' allergies, patients' names, addresses, telephone numbers, and care hours scheduled for each patient, and the names and phone numbers of their pharmacies.

The nurses could record each action they take with each patient, as they do it via the keyboard. At the end of each day this information would be transmitted electronically into the PC at the nursing station for billing, report generation, and such other administrative needs as complying with government regulations. Preliminary indications are that visiting nurses will be able to increase their patient load from the current five per day to as many as seven. Clearly this example manifests a potentially major application of the benefits to all involved parties, from provider of technology through to patient, of this type of system.

InfoMed is also evaluating a patient computer that could prompt and record self-monitoring regimens, self-administered medications, outpatient treatment appointments, and home care nursing visits. Each patient's emergency medical information file could be immediately accessible to anyone. A personal medical history, accessible only to the patient and doctor, could also be included.

## 2.8   INSURANCE SALES AID

Connecticut Mutual Insurance Co., other insurance companies and SCI have jointly tested UltraSmart Card as a sales aid for insurance agents. UltraSmart Card type of technology is capable of instantaneously calculating insurance rates utilising the most current insurance tables and can also simultaneously store the actual rates of several customers. The customer data can be quoted and can be brought back to the field office and transferred to a PC or mainframe for further processing or record keeping purposes. Insurance tables may be updated and downloaded at any time by inserting the UltraSmart Card into a card reader which has been connected to a PC at the home office.

To use this type of card, agents enter their PIN which protects sensitive data from unauthorised access. They specify the type of insurance required for quotation by the client, existing or prospective, from a menu on the card — one card can handle calculations for several different kinds of insurance. They choose an appropriate insurance plan from a menu. They are prompted to enter the sex and age of the client and whether they smoke. Then they enter the face amount of the policy. The card calculates and displays the standard annual premium and the preferred annual premium. It displays the guaranteed cash value of the policy, total death benefit at age 65, and such options as dividends, surrender values and available load amount for whatever year is specified. This process enables information to be delivered immediately in either home or office, creates better client service, and creates more revenues for both the agent and his company. Other insurance carriers have indicated interest in following these types of applications.

**Fig. 2.5** One side of the UltraSmart Card prototype designed to deliver travel and related financial services.

## 2.9  TRAVEL AND RELATED FINANCIAL SERVICES

Via both its US and UK offices and marketing/engineering facilities, SCI and Thomas Cook (Peterborough, UK) in 1987, signed a very significant development contract for prototypes of the first advanced (UltraSmart Card) smart card for the delivery of travel and related financial services (Fig. 2.5). The card provides its built-in keypad and display screen to re- cord and monitor for travellers their spending on such items as expenses, and electronic travellers cheques or electronic cash, plus itinerary plan- ning, airline and hotel preferences and corporate/organisation travel policies.

Peter Middleton, Chief Executive of Thomas Cook described the appli- cation: 'Our commitment to this major project reflects our determination to deliver travel and financial services with greater efficiency at reduced costs. While these services will enhance our core business, they will also simplify travel complexities before, during and after journeys. 'For ex- ample, once back in the office the traveller can insert his or her card into a smart card reader attached to a PC and immediately receive a hard copy print-out of incurred expenses collated by category.'

The Cook's prototypes were delivered in 1989 and are expected to be the forerunner of production numbers for future major use by Cook and other related organisations worldwide.

## 2.10   FUTURE DEVELOPMENT

As exciting as these developments are, they are only the beginning for smart card technology in the world. Back in 1982, this writer developed a probable timetable for smart card implementation in North America that so far has proven to be valid and which I believe will continue to be accurate. It projected that in 1983–1985 the first early tests would take place in the US, and that happened. It also predicted customisation of the technology and development of the early market, both of which happened. In the short-range future – 1986–1988 – it predicted the appearance of the first substantial applications and the beginnings of proliferation of uses, and that occurred on schedule. In the mid-range future – 1989–1992 – it projected the development of more extensive applications, initiation of mass manufacturing, and the creation of a major industry. With mass production, smart cards can be expected to cost anywhere from 50 cents to $50 each, depending on what they have built into them and what software is required in them. And in the longer-range future – 1992–2000 – my timetable projected worldwide, applications – essentially smart cards in many aspects of human and organisational life.

This does not mean that every human being will have a smart card; however, for a very large number of the world's populace smart cards are likely to manifest themselves in all kinds, shapes, and sizes. This is because they are the logical extension of distributed processing – from paper to desk top computer to pocket or purse PC. We will probably never totally eliminate cash or paper, but we will truncate the use of both. The reason all this will happen is that the smart card technology – and in future applications the UltraSmart Card/SuperCard form of it – will meet the immediate and daily needs of a large number of present and potential users, whether in a city or tribal wilderness, to access government entitlements.

Potential smart card users today are looking for solutions to the increasingly vital problems of managing information effectively and conveniently. Bank cheque customers want a cheque book that automatically balances, that guarantees that no one can forge a cheque and is always available when wanted. Medical patients want an easy way to keep track of their diet without having to carry around a notebook and pencil and constantly scribble down information and add up lists of calories. Government wants to provide human services to its citizens more effectively and less expensively.

However, vendors/manufacturers of smart cards must learn to conform to customer requirements which go beyond selling products. Virtually all these actual and potential users want most or all of five criteria to be met in a system context:

(1) applicability to their needs and conformity with existing operations environments and/or systems;

(2) easy delivery and accessibility of information;
(3) security, which has been a hallmark of smart card from its inception;
(4) flexibility, which the technology provides inclusive of such improvements as logical data file storage; and
(5) economic practicality and viability.

As new generations of the technology appear with more improvements in processor speed and memory capacity, these devices will inevitably find many more applications in the government, medical, industrial, petroleum, retail, auto rental, insurance and many other industries. They will be adopted as handy devices for accessing information on networks – a skill that is already becoming vital to many professionals. They will also likely replace floppy discs because of security as well as increased memory capabilities. Transportation companies will use them instead of tokens and public phones and vending machines will use them instead of coins, but most applications probably have yet to be identified and developed.

Much of all this is happening now in France and Western Europe and in Japan. And yes, eventually even MasterCard and Visa will use them in one form or another, pressured by their acceptance by such a broad cross-section of the marketplace. In the US, as in other countries, the entire thrust of technology today is towards increasing computerisation, towards more effective control and dissemination of information in cars and appliances, home security and energy control, as well as throughout offices. This makes the replacement of the present, 40-year-old magnetic stripe technology with the new 'smart' digital technology and its vastly superior information/service delivery capabilities inevitable both in everyday applications, a host of new ones, and eventually in the 1990s in traditional credit card applications as well.

There is a growing perception in the US and elsewhere among independent research groups, technologists, smart card users and potential users that the smart card will change the way we do things – that smart card technology will change the world. It is likely, despite a slower start, to take hold most broadly and profoundly in the US. It is, I believe, certain to be the major disseminated information technology in the world as we enter the 21st century.

**SUMMARY –**
**AN AMERICAN PERSPECTIVE CHRONOLOGY**
**OF THE HISTORY OF SMART CARD TECHNOLOGY**

| | |
|---|---|
| 1970 | Dr Kunitaka Arimura files first basic smart card patent in Tokyo (Japan only). |
| 1974 | Roland Moreno (A French financial/technology journalist) invents and files first broad based smart card patents in |

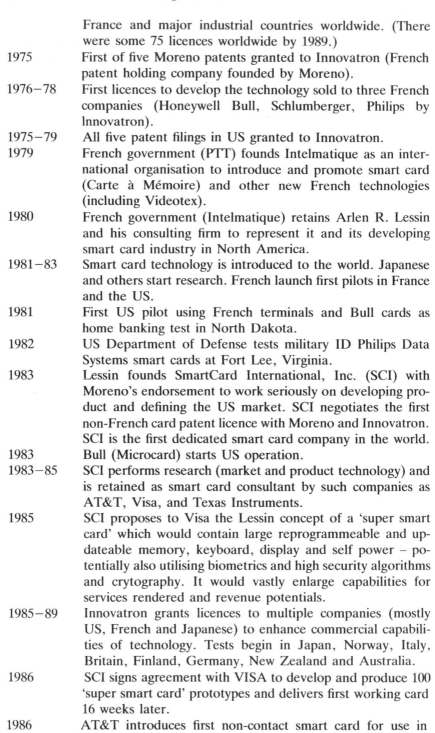

France and major industrial countries worldwide. (There were some 75 licences worldwide by 1989.)

1975        First of five Moreno patents granted to Innovatron (French patent holding company founded by Moreno).

1976–78     First licences to develop the technology sold to three French companies (Honeywell Bull, Schlumberger, Philips by Innovatron).

1975–79     All five patent filings in US granted to Innovatron.

1979        French government (PTT) founds Intelmatique as an international organisation to introduce and promote smart card (Carte à Mémoire) and other new French technologies (including Videotex).

1980        French government (Intelmatique) retains Arlen R. Lessin and his consulting firm to represent it and its developing smart card industry in North America.

1981–83     Smart card technology is introduced to the world. Japanese and others start research. French launch first pilots in France and the US.

1981        First US pilot using French terminals and Bull cards as home banking test in North Dakota.

1982        US Department of Defense tests military ID Philips Data Systems smart cards at Fort Lee, Virginia.

1983        Lessin founds SmartCard International, Inc. (SCI) with Moreno's endorsement to work seriously on developing product and defining the US market. SCI negotiates the first non-French card patent licence with Moreno and Innovatron. SCI is the first dedicated smart card company in the world.

1983        Bull (Microcard) starts US operation.

1983–85     SCI performs research (market and product technology) and is retained as smart card consultant by such companies as AT&T, Visa, and Texas Instruments.

1985        SCI proposes to Visa the Lessin concept of a 'super smart card' which would contain large reprogrammeable and updateable memory, keyboard, display and self power – potentially also utilising biometrics and high security algorithms and crytography. It would vastly enlarge capabilities for services rendered and revenue potentials.

1985–89     Innovatron grants licences to multiple companies (mostly US, French and Japanese) to enhance commercial capabilities of technology. Tests begin in Japan, Norway, Italy, Britain, Finland, Germany, New Zealand and Australia.

1986        SCI signs agreement with VISA to develop and produce 100 'super smart card' prototypes and delivers first working card 16 weeks later.

1986        AT&T introduces first non-contact smart card for use in public phones and other potential applications.

1986    Bull Microcard awarded conventional smart card contract by US Department of Agriculture for peanut farmer application.

1986    SCI completes public securities offering and becomes first full spectrum publicly funded and traded smart card company in the world.

1987    SCI files US and international patents on its UltraSmart Card and also proprietary software and hardware technology and operating system software.

1987    PC3 issued smart card patent.

1987    Multiple tests/trials begin internationally.

- France commits to reaching 32 million cards by the end of 1988 – including bank cards.

1987    Smart Card Applications and Technologies (SCAT) organisation founded and holds first conference in US.

- US interest increases significantly, especially from Federal and State governments.
- SCI completes SuperCard prototype deliveries to Visa.
- Bull (Microcard) signs contract for US Marines security card. Some small US companies market smart card test contracts.
- Schlumberger contracts to supply smart cards to State of Michigan for job placement ID and other services.

1988    Significant market events:
- SCAT breeds ESCAT (European Smart Card Applications and Technologies) meetings (1988 and 1989) in Helsinki. Also, ASCAT (Asian Smart Card Applications and Technologies) is being organised and its first conference is set for Tokyo in 1990.
- Market develops substantively in other than banking sector.
- First US UltraSmart Card technology patent is issued to Lessin and other SCI associates (patent assigned to SCI).
- SCI signs contract with Hawaii state government to develop smart card model program to render human services to residents (SCI UltraSmart Card type).
- Kodak-Pathé contracts for smart card transportation application (SCI UltraSmart Card type).
- AVIS contracts for smart card auto leasing service tracking application (SCI UltraSmart Card type).
- Schlumberger, PC3, SCI approved by US National Security Agency (NSA) for security application smart cards.

1989        • 'SmartStart' for secure access to PCs developed (SCI).
            • Trials of first supermarket POS smart card applications in
              the US (PC3).
            • US Veteran's Administration issues request for proposal
              for smart cards containing medical profiles of infirm
              veterans.
            • US Department of Agriculture issues request for proposal
              for smart cards for 1990 food stamp delivery trial.
            • State of Michigan broadens its smart 'opportunity card'
              program for state residents.
            • AT&T and Olivetti reach agreements for Olivetti to pur-
              chase and issue AT&T card for mass transportation appli-
              cation.
            • Lessin leaves SCI to found Lessin Technology Group,
              Inc. (LTG) with mandate to provide smart card and
              related technology consultative and systems services
              world-wide. LTG is first company to address the bringing
              technology 'from concept to practice'.
            • LTG establishes European office in UK. Roy D. Bright,
              former head of Intelmatique (France), is Managing
              Director.
            • LTG and SCAT organisation agree to hold executive
              briefings in US and overseas under the aegis of the Inter-
              national Smart Card Institute (ISCI). Its purpose is to
              disseminate unbiased information about the technology
              and its realistic applications. The President of ISCI is
              Lessin.
            • Trials and plans for applications profliferate in wider
              ranges, including newly identified areas.

1990–93     Watershed years for the acceptance internationally of smart
            card and related technologies.

Chapter 3

# A Contactless Smart Card and its Applications

JOHN McCRINDLE

(Marconi Research Centre)

*We discover a very practical alternative to the contacts of the original smart card.*

## 3.1 INTRODUCTION

The late 1940s and 1950s saw the introduction and growth in numbers of the plastic card. By the early 1970s their use was widespread in areas such as finance, travel and entertainment. At the same time, computers developed rapidly, culminating with the introduction of the microprocessor in 1971. This device was to appear in all forms of equipment and consumer products from spacecraft to washing machines.

The idea of combining these two products by embedding a microprocessor in a plastic card was conceived in the mid 1970s. Today, this product, the smart card, looks like having a greater impact than the plastic card and microprocessor combined. During the next decade it will pervade all areas of our lives as it is used in finance, medicine, the armed services and telecommunications, and in more far-reaching applications such as passport replacements. It is also set to change fundamental ways of working that have been carried out for centuries as it becomes a direct electronic replacement for money.

The first applications of smart cards took place in France using cards which had surface contacts for powering up the electronics within the card, and for communications. This type of card requires a read/write unit with a slot into which the card has to be presented correctly and forced home. Operational reliability is very much dependent on the careful treatment of both the card and the read/write unit which are prone to wear, contamination and damage (deliberate or otherwise). In many cases, before a new product is widely accepted it is superseded by a new generation which becomes the internationally recognised product. The GEC ic Card is one of a new type of smart card which is now emerging. As a result of research which began in 1983, GEC Card Technology has overcome the deficiencies of the earlier contact type smart card with the introduction of a revolutionary new concept — a

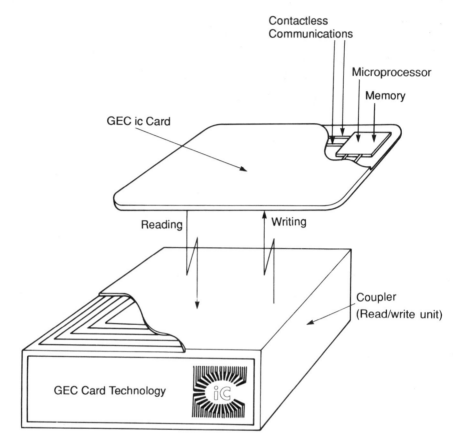

**Fig. 3.1** The GEC intelligent contactless Card and Coupler (read/write) unit.

smart card which can communicate via a unique contactless interface (Figure 3.1).

## 3.2   THE GEC INTELLIGENT CONTACTLESS (ic) CARD

The GEC intelligent contactless (ic) Card has been designed to be a conveniently sized token that is both computationally powerful and very secure. Figure 3.2 shows the main elements of the card, and read/write unit, or coupler as it is known in its simplest form.

Sealed into the card is a microprocessor, memory and contactless interface. In practice the microprocessor and memory are on one chip.

The microprocessor is the intelligence within the card. When activated it executes instructions programmed into the memory. It can perform

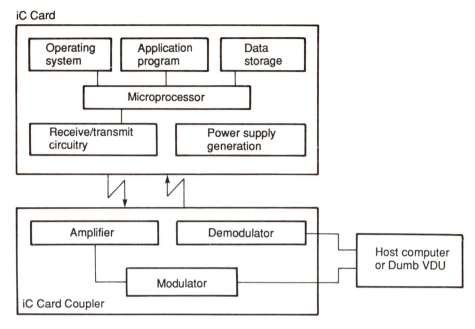

**Fig. 3.2** Block diagram of the elements of the GEC ic Card and read/write unit.

calculations, manipulate data, carry out security checks to ensure that data stored in the card is not revealed to unauthorised persons, and it can scramble and descramble data sent to and from it. It can also compress data that is to be stored in the card and expand data that is to be transmitted to the outside world.

The memory in the ic Card is functionally partitioned into three areas. One area is taken up by the operating system which controls those functions such as data transmission and reading and writing of memory, which are fundamental to the card's operation. The operating system is embedded in the chip at the manufacturing stage and cannot be altered. In a second area resides the application program, that is the segment of executable instructions which defines in full the behaviour of the card within the environment of the particular application to which the card is to be put. This program is loaded by an operating system function which may be irreversibly disabled. The remainder of the memory is completely under control of the application program, and is generally used as read/write data storage.

The power for the electronics and communications link is provided by the contactless interface. Built into the ic Card is a small coil of wire which develops a voltage across it when the card is in the presence of an inductive radio frequency field. The voltage is rectified and regulated on the card to provide a steady power supply for the electronic elements. Also built into the card is a reset circuit, designed to activate the card when it is in the correct field. The RF field is also used as a bidirectional data path. Data is sent to the card in serial fashion by shifting the frequency of the RF carrier and this is decoded in the card.

The coupler, which is linked to the host system by a standard (RS232) serial data line, acts as an interface for the data flowing between the card and host system. It produces the RF field, the frequency of which is determined by the state of data from the host, which is decoded in the card. The coupler also contains amplitude modulation (am) detection circuitry for converting information received from the card into a serial data stream. The data rate can be anything from 300 to 9600 baud with even or odd parity, if desired, and any number of stop bits. These characteristics are fully under control of the card software. The field strength from the coupler is sufficient to allow card operation up to 20 mm from the coupler but falls off rapidly after this distance. As its name suggests, the coupler merely acts as the means by which the card communicates to and from the other system components.

These technical features provide the GEC ic Card with unique benefits. The contactless interface means that the life of the card is much longer than conventional smart cards which are susceptible to contamination, damage and wear of surface contacts. In use the card can be placed in any orientation on the surface of the coupler, or the coupler can be mounted underneath any non-metallic working surface. This is very useful in banks and at retailing checkouts, keeping counters free of equipment. Unlike conventional card reading mechanisms, the couplers are sealed electronic units with no moving mechanical parts. Thus they are able to withstand harsh environments and severe treatment yet remain very reliable, require minimal maintenance and can be easily incorporated into existing equipment. As both card and coupler electronics and connections are totally sealed (Figure 3.3) they are capable of working in wet and dirty environments.

## 3.3  SECURITY FEATURES

Many of the applications of the smart card require security somewhere in the system whether it is for the protection of sensitive data, protection against eavesdropping or protection against illegal use. The GEC ic Card has a range of security features impossible in magnetic stripe cards and unrivalled by many computing based products. It offers:

(1) Protection against the manufacture of fake cards because of the highly complex technical nature of the card and high cost of manufac-

**Fig. 3.3** The GEC ic Card and Coupler electronics are totally sealed making them ideal for use in harsh environments.

turing equipment which acts as a deterrent to all but the largest criminal organisations.

(2) Protection against easy access to the electronics, and hence data storage area, by complete encapsulation of the electronics.

(3) Protection against the probing of data lines between microprocessor and memory by incorporating both the elements on a single micro-electronics chip.

(4) Protection of the application program through the ability to 'blow' a software fuse thereby destroying the means by which the card can reload a new program.

(5) Sumcheck protection against the alteration of memory contents.

(6) Protection against altering and adding to the dialogue between the card and a terminal by authentication software specifically designed for the card.

(7) Protection against using recorded dialogue to establish authentic communication and against rerouting messages by verification software specifically designed for the card.

(8) Protection, through encryption, against deciphering dialogue between the card and terminal.

(9) Positive personal identification of the card holder by comparison of a personal characteristic (e.g. signature, fingerprint, facial features)

of the legitimate card holder stored on the card with the same
feature of the person presenting the card at an access point. The
comparison can be carried out within the card, thus maintaining
complete secrecy of reference data.

(10) Protection, by the card invalidating itself when repeated attempts
are made to gain access by continued keying in of possible personal
identification numbers or forging of signatures.

In totality, the security offered by the ic Card is virtually unrivalled by
any other low cost computing based product.


## 3.4  APPLICATIONS

Applications for the smart card can be divided broadly into three categories: data carrier, where the card is used as a convenient portable and
secure means for storing data; conditional access, where the card is used
as a secure means of identifying the holders entitlement to gain access to
a site, a computer, a software package or a service; and financial, where
the card is used to replace credit cards, cheque books or money. Each
card is by no means restricted to one application only. A card can
accommodate several different functions spanning all three categories.
For instance, one card could be used to hold medical data, provide access
to a computer system and act in a financial capacity.

As a data carrier the card has many applications in the medical field.
Used as a general medical card, the ic Card could contain such information
as the holder's address, date of birth, name and address of his/her doctor,
allergies, recent medical history, serious complaints, drugs being taken
and donor wishes. The card could be carried by the individual and in the
case of an emergency, for example the holder collapsing in a street or
being involved in a road accident, would provide immediate medical
information to the ambulance crew (Figure 3.4) or the doctor in a hospital
casualty department. The speed with which vital information would be
available could well save lives. The card is also particularly suited to
patients requiring regular treatment or regular monitoring e.g. diabetics,
dialysis patients. In these applications the card allows key information to
be provided easily and quickly to the doctor at each appointment and
data can be easily added to the card.

Military applications include electronic identity tags for servicemen and
women. The card can contain details of the holder, service records,
medical history, entitlements etc. The card is particularly suitable for data
logging. At remote or unattended sites it could be used to record temperature, events etc. Periodically it could be collected and returned to a
central point for the logged information to be read off the card.

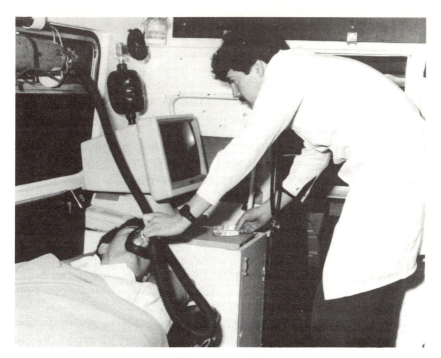

**Fig. 3.4** The GEC ic Card could contain medical details about the holder. In the case of an emergency it could provide vital medical information to an ambulance crew or doctor in a hospital casualty department.

As a maintenance record, the card could be conveniently attached to equipment. The paperwork that goes with military and high value industrial equipment is often considerable. The smart card provides an easily updatable compact way of storing such data.

There are many industrial applications for the card. For example, it could be used to program computer-numerically-controlled (CNC) machines replacing punched cards or magnetic tapes. Alternatively a card could be used to store a record, for monitoring purposes, of the progress of manufactured components throughout stages of their manufacture. In the automobile industry, such a card might subsequently form the basis of a vehicle's servicing record.

In the airline field the card could be used as an electronic ticket with a complete analysis of the passenger's preference for 'smoking' or 'non-smoking' seat, as well as dietary needs. For regular travellers it could log the number of trips flown with a particular airline to give a free or reduced fare flight after a number of trips have been made.

In the area of secure access, the card can act as an electronic key to control access of personnel to facilities where sensitive work is carried out or data is held. The most common type of security access devices are keys, badges and magnetic cards. These all suffer from the same basic drawbacks that they can be easily duplicated and when stolen or passed on to someone else, either wilfully or through coercion, they can allow entry because there is no link with the person to whom the device was issued. The ic Card overcomes these weaknesses because it is very difficult to reproduce and has the capability of storing a digitised personal character-istic of the owner (e.g. fingerprint). With suitable verification equipment, this data can be used at the point of entry to identify whether the cardholder is the legitimate owner of the card. The card also has the benefit that it can easily be individually personalised to allow access to only certain facilities depending on the security clearance of the card holder. Additionally, as the cardholder progresses through a security system, a log of the person's movements can be stored on his card as a security audit trail.

Computers often hold sensitive information and access to this information has to be controlled. The smart card offers a solution. It can hold a cryptographic key to allow access to various areas of a database depending on the card holder's level of authority.

The smart card also offers a solution to the problem of unauthorised copying of software. By storing a key part of a software program in the card, the complete program will only be able to run with the smart card present.

Direct Broadcasting by Satellite (DBS) and Cable Television are going to become more widespread in future years. The smart card offers a means for payment and the key for reception. Customers will be able to purchase an ic Card that will provide the necessary key to unscramble the picture. Cards and decoding equipment could be supplied through TV rental companies. After, for instance, an interval of one month the key required to decode the signal can be changed so that the user has to re-turn to the rental shop to have, upon payment, the card updated with the new key. Viewing time statistics could be simultaneously collected.

Banks' major clients can use the ic Card as the key to secure access of the bank's mainframe computers for corporate cash management. The card is a secure token for individual companies to access their bank accounts and financial services from remote personal computers on their own premises. This service could later be extended into home banking.

In the general financial area the card can be used in a number of ways. It can be used to replace the cheque book. At a point of sale the smart card has the capability to compare the card holder's personal identity entered by means of a four digit number, or characteristics of a digitised signature, with a secretly held reference in the card. A correct comparison will then allow the automatic transfer of funds from the purchaser's bank account to the retailer's bank account.

**Fig. 3.5** The GEC ic Card can be used as the means for paying for goods at a retail outlet.

The card can also be used as an electronic wallet replacing cash. Here the card will have prepaid amounts which can be used for payment of low value purchases in shops (Fig. 3.5), at vending machines and car park entry points by the automatic deduction of the appropriate amount. The payment made will be held securely within the vending machine, probably on another smart card, for subsequent reconciliation.

As an electronic token, the card is equivalent to the electronic wallet but instead of cash, holds units of consumption such as electric and gas units and telephone charge units. In applications such as these the card could also provide additional facilities. In the case of the electricity/gas card it could monitor and store when units are being used; information which could be extracted from the card when next the token value is replenished. In the case of the telephone card it could also hold telephone numbers for speed dialling.

In the longer term the card could be used as a social services card carrying individuals' child allowance, pension entitlement or social security entitlement. It could be used as a driving licence, tax disc and log book, readable electronically through the car windscreen. One day it is envisaged there could even be an 'electronic' passport where the card is simply laid upon the counter of immigration control to securely validate the holder and expedite the immigration and visa checking process.

## 3.5  THE FUTURE

There seems little doubt that the smart card will start to have a major impact in the early 1990s. It is already being extensively adopted in France. In Japan most of the major electronics companies are rapidly developing smart cards and a number of trials are underway. In the USA, potentially the largest world market for smart cards with a reported 825 million plastic magnetic stripe credit and debit cards already in existence, major trials and implementations are already beginning or are expected soon. In the UK the GEC ic Card is being used in a number of areas including the first trial of smart cards by a UK bank for financial applications. Since its introduction, the GEC ic Card has attracted worldwide interest and orders have been received from the USA, Europe and Australasia. It is set to take a major share of the emerging smart card market.

Chapter 4

# Low Frequency Radio Tags and their Applications

## JOHN FALK

(Contag International Ltd)

*Radio tags are well established cousins of the smart card.*

## 4.1 INTRODUCTION

Radio tags are a logical development of the bar code industry. The success of bar codes had demonstrated a growing and universal need to identify items quickly and reliably. As a technology, bar codes are easy to use and have the added attraction that the labels are very low cost. In a considerable number of applications therefore, particularly where the label count is high, bar codes will continue to be the dominant technology for the foreseeable future.

Despite their obvious advantages not all identification problems may be solved with bar codes. In fact on closer study they exhibit a number of limitations which restrict their universal use. For example the bar code must be in direct line of sight with a reader in order to be identified. Thus dirt, condensation, misorientation and misalignment can all contribute to misreads. Furthermore, except for very expensive readers, identification must take place at a predefined distance between reader and bar code. Another restriction is that the amount of information that can be contained on a bar code is strictly limited by its size and this information, once printed, cannot be changed.

These limitations have led to the development of alternative technologies, one of which has been the emergence of the radio tag. Such devices fall broadly into two main categories. There are high frequency tag systems which operate generally in the microwave band. They tend to hold large addressable memories and are usually relatively expensive. There is also a growing army of low frequency tag systems. These operate predominantly in the inductive communication band between 10–150 kHz. It is these low frequency systems which form the subject of this chapter.

## 4.2  ELEMENTS OF A CODED TAG SYSTEM

Before going into a more detailed study of LF tags, it will be useful to introduce the reader to the basic elements of the system. A typical system is illustrated in Figure 4.1 and comprises the following:

*Coded Tag*    This device will be attached to each object, person or vehicle which is to be identified. The coded tag is small in size (typically $40 \times 40 \times 10$ mm) and built to withstand rugged use.

*Programmer*    The function of the programmer is to enter a predefined code into each tag. This code may consist of an identity number, or may

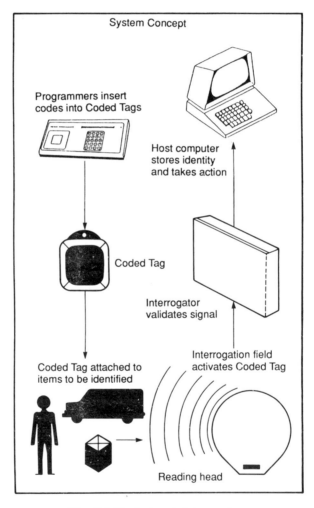

**Fig. 4.1** Typical coded tag system.

comprise both identity and data. The programmer can also interrogate a tag. It may therefore be used in some specific applications as a read/write unit.

*Reading head*    This unit combines the transmit and receive antennae which interrogate a tag. The reading head is connected to the interrogator vià a feeder cable.

*Interrogator*    The interrogator generates the interrogate signal for the transmit aerial. It amplifies the response from the tag which is then validated before being output as data.

*Host*    The host will usually be a computer or PLC. It receives valid data from the interrogator and takes appropriate action.

Figure 4.1 only attempts to describe the basic elements of a coded tag system. In practice a large installation may comprise a large population of reading heads, interrogators and programmers.

## 4.3  BENEFITS OF LOW FREQUENCY

Perhaps it is not surprising that LF tag systems had a harder struggle for acceptance than their high frequency cousins. At first sight the choice of low frequency does not look reassuring. Antennae must inevitably use coils comprising many turns and frequently must incorporate ferrites. Data rates will be relatively slow and furthermore ambient noise levels from electric motors, VDUs, switched mode power supplies etc. predominate in the LF band. It is only when one takes a serious look at the advantages of an inductive system that the real benefits become apparent. Since these benefits are fundamental to the technology we will cover them in some detail.

A low frequency system enables the designer to use a single CMOS device within his tag. This offers a number of advantages. Since CMOS circuits have a very high input impedence they may be directly coupled to the tuned input antenna without adversely affecting the Q. It is also a relatively easy task to bias the CMOS input amplifier so that it operates at a predefined threshold. By this means it is possible to achieve quite acceptable input sensitivities without the need to resort to sophisticated input amplifiers.

CMOS circuits also are well known for their low consumption of current. Provided the input amplifier bias level has been correctly selected, the quiescent current of a CMOS device is negligible. This is particularly valuable where the tag contains a volatile memory and therefore incorporates some internal power source. Even in their operating condition CMOS devices still draw only small levels of current. This again makes them an ideal choice for both active or passive tag systems.

A second benefit of low frequency systems is that the communication medium is predominantly magnetic and not electromagnetic. Thus all of the basic equations of magnetic fields about a coil apply. The most important of these is the relationship in which

$$H = \frac{I\, a^2\, (N)}{2(a^2 + z)}$$

where H = the field generated by the coil in AT/m
     I  = current flowing through the coil
    N = number of turns in the coil
    a  = radius of the coil in metres
    z  = the distance in metres along the axis of the coil at which the field H exists

This is particularly useful since it shows that beyond the immediate near field of a coil, the intensity decays as an inverse cube of the distance. Such a relationship is ideal for an identification system. A reading head will have a high field level in its immediate proximity but a very low level in the far field (see Figure 4.2). Such an arrangement ensures that tags which are being read are exposed to a high field as they pass in front of the antenna. However, other tags which may be in the vicinity will not be inadvertently activated.

A further side benefit of this effect is that interference generated by electric and electronic devices is also subject to the inverse cube law. Thus the simplest technique for overcoming interference from a neighbouring source is to separate by a short amount the distance between them. Of course this may not always be possible, in which case more sophisticated methods will be necessary.

Magnetic fields will also penetrate dense materials with very little loss in field strength. This is convenient since it permits tags to be read when situated within or behind a solid object. Equally it is possible to read tags while immersed in a fluid without any significant loss in performance.

Low frequency magnetic fields will not of course penetrate electrically conductive materials. However, the field is able to 'go around' metal objects in a way which gives the effect that a tag on the far side is being read through it. In fact magnetic fields show a quite remarkable ability to pass through even the smallest of apertures. The author on one occasion arranged a demonstration in which a tag was read inside a tobacco tin. In this case the signal passed through the small air gap between the lid and case.

As a consequence of the inverse cube law, low frequency systems are relatively immune to the misorientation of the tag with respect to the reader. This point can be demonstrated by the following simple analysis. The antenna in an LF system will consist of a coil comprising one or more

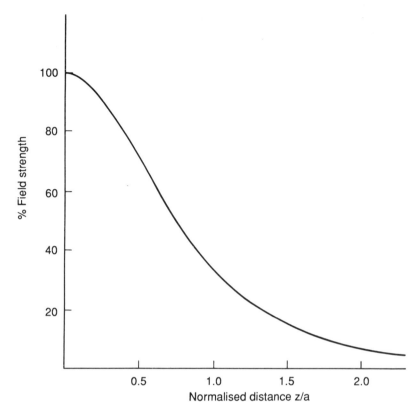

**Fig. 4.2** Magnetic field strength around a coil.

turns. Where a transmit and receive coil are placed in the same plane the coupling between them will be a maximum. As the receive coil is rotated with respect to the transmit coil, the coupling is reduced by the cosine of the angle of rotation. The table at Figure 4.3 shows the percentage coupling for a number of typical values for the angle of rotation 0. However, for a given input threshold level the reduction in range of a tag from the transmit coil will be the inverse cube of the cosine of the angle 0. Thus a quite significant rotation from optimum will result in only a modest reduction in range. This holds true for values of 0 up to 60°. Thereafter the range will drop away as 0 tends to 90° where the range is theoretically zero. In practice as the tag passes across the reading head even in worst orientation the curvature at the edge of the field ensures a reasonable level of coupling. The cumulative effect of the above is to allow a tag to be read acceptably in almost all orientations.

A final point worth mentioning is the matter of international PTT approvals. This is an area which can be easily overlooked, yet worldwide

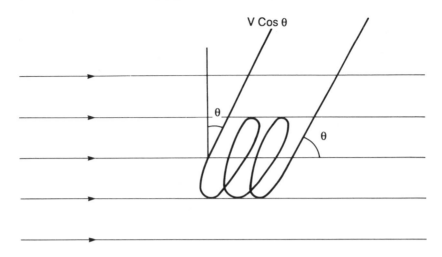

| Angle θ | % Coupling | % Reduction in range |
|---|---|---|
| 0 | 100 | 0 |
| 45 | 70 | 11 |
| 60 | 50 | 21 |
| 75 | 26 | 37 |
| 85 | 9 | 56 |
| 89 | 2 | 74 |
| 90 | 0 | 100 |

**Fig. 4.3** Reduction in range with rotation.

controls are growing increasingly more stringent. Due to the declining availability of frequencies in the shortwave and microwave bands, allocations granted in one country may not be available in others. Without advance planning this can turn out to be an expensive setback in a marketing programme where product development is already complete. By comparison with other parts of the radio spectrum, regulations in the inductive communication band are far less stringent. In fact a number of countries do not require any form of PTT approval for devices operating below 150 kHz. The choice of low frequency can therefore make a significant saving in a company's approval costs.

## 4.4  PRINCIPLE OF OPERATION

With practically all RF tag systems the tag operates as a transponder. Such a device will only be activated when subjected to an external inter-

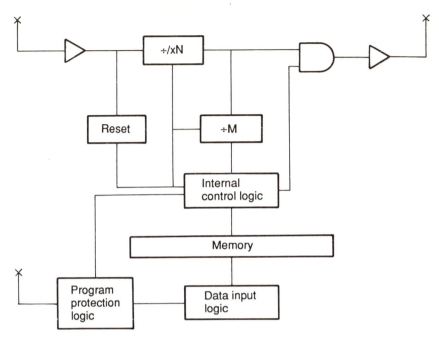

**Fig. 4.4** Outline schematic of a typical coded tag.

rogation field. This mode of operation is ideal for a tag system. For most of its life a tag will remain in a dormant condition. However, when it is brought within the field from a reading head it will immediately switch to its active state and commence to respond.

The interrogating signal frequently serves two functions. Not only does it activate the tag but also it provides the clock frequency which controls its internal operation. A typical tag system might be configured as in Figure 4.4.

The circuit comprises two analogue inputs. One of these is tuned to the interrogating frequency while the second is set to a different frequency to receive input data. The interrogation frequency is fed to a frequency divider or multiplier which generates an output signal. The interrogation frequency is also used to time the internal logic. This neatly ensures synchronisation of any input or output data. Additionally it avoids the need for an internal oscillator within the transponder.

The internal logic will validate and accept input data and control its storage in memory. It also accesses from memory as required output data which is passed to the driver stage. The data is used to modulate the output carrier using either phase or amplitude modulation.

A reset circuit is incorporated to ensure synchronisation of the internal logic. In the quiescent state this ensures that the memory is protected and all flip flops are held in a reset condition. As soon as the tag is brought within range of an interrogating signal, the reset line is lifted and the internal logic will commence to operate.

In a number of systems additional security is incorporated to protect the memory. This is done to prevent either accidental or deliberate corruption of the database. The memory protection circuit is contained within the internal logic and will validate a password or sequence from the data input before giving access to the memory.

We have not yet covered the synchronisation of messages between the tag and the interrogator. A number of different techniques are possible. For example a unique preamble in the data stream may be used to signal the start of each message. Another technique is to pulse modulate the interrogate signal. This enables the reset line in the transponder to be lifted at the start of each new interrogation pulse. Thus an output message from the tag is time related to the start of each interrogation cycle.

## 4.5   TAG CONSTRUCTION

The heart of any tag is its custom microchip. This will be fabricated in CMOS and contain both the analogue amplifiers and digital elements of the tag. The requirement for both analogue and digital elements on a single die will inevitably stretch the abilities of the chip design house to its limits. On the one hand the digital elements of the circuit must perform sufficiently fast within acceptable limits of data skew. Simultaneously the input amplifiers must operate at a high level of input sensitivity while drawing minimal current. All of this must be achieved at minimum cost while ensuring acceptable tolerances between wafer batches. Small wonder that the gleam in the chip manufacturer's eye quickly fades as the reality of the requirement sinks home. Not surprisingly the specification phase for such a custom device is likely to be protracted with considerable debate before final agreement is reached.

We should not leave the microchip without covering the subject of memory size. It is convenient to subdivide coded tags' capacity into a number of levels as represented in Figure 4.5. For completeness the single bit presence sensing tag, as used for example in anti-shoplifting applications, is also included. It will quickly be apparent that capacity heavily dictates the use to which each level is put. LF tag systems are available with memory capacities which range from 8 bits to low thousands of kilobits. Not surprisingly low memory tags and their associated control equipment are significantly less expensive than the high memory versions. A low memory is useful in applications where an identity number only is required. In such situations it is frequently unnecessary to reprogramme the tag. Where this applies a passive system may offer the most cost-effective solution.

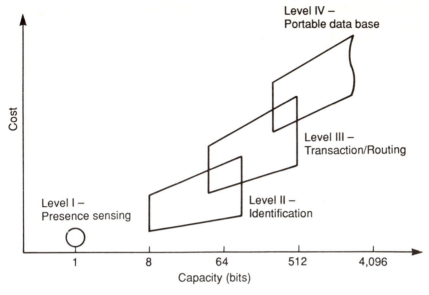

**Fig. 4.5** RF coded tag systems – capacity.

As memory size increases, the opportunity to combine data with identity becomes a possibility. The data may take the form of flags which occupy a predefined section of the memory. These flags might be set, for example, to indicate the successful completion of different stages in a production process. Alternatively the data may be in the form of a number which relates via a look up table in the host to a sequence of actions. Medium sized memory systems may be loosely classified as comprising between 32 and 256 bits. They are characterised by the fact that the data is usually associated with a look up table.

High memory systems are those which comprise memories greater than 256 bits. They are distinguished by a number of features. Firstly the memory is arranged in 8 bit bytes with each byte being individually addressable. This is clearly essential to avoid the need to write to or read the entire tag memory at each read/write head. Secondly the large memory size makes it possible for the data in the tag to be entered in full in ASCII thus avoiding the need for external look up tables. This is a real benefit in applications where there would otherwise be a frequent need to modify a distributed database. On the other hand, each work station must contain the necessary software to enable it to access its own data requirements from the tag. Also if the quantity of data to be transferred at a reading head is large, the time taken to pass the data may be significant. It will be apparent therefore that considerable work is involved in the systems design and installation of a high memory device. For this reason their uses are currently limited to large production lines where the cost can be fully justified.

Coded tags may be either active (battery powered) or passive (battery-less). During interrogation, the passive tag draws its power from the interrogation pulse. This is used to operate the microchip and drive the output data signal. Such systems can work well at short range. However, at extended ranges or with misorientation the output signal is reduced. This effectively limits the market for passive systems to use at short ranges and may restrict their ability to operate in electrically noisy environments. Active tags usually contain a small lithium cell which maintains the memory during the quiescent state and powers the device to give a strong output signal during interrogation. Lithium cells are particularly useful as a power source because of their very high power density and long shelf life. They will also operate effectively over a wide temperature range with typical operational limits from 75°C down to −30°C. Although lithium cells are capable of delivering only small current levels, this is not an obstacle to their use in a tag where minimal current drain is a basic design criterion. Most active tags will provide typically 7 years of operation under conditions of high use. Thus in practice the service life of the tag is likely to be exceeded long before the cell is discharged. However, as an additional precaution most active tags incorporate a battery sense circuit within the microchip. This will check the cell voltage each time the tag is interrogated. In the event that the battery voltage drops below a predefined level, a flag will be set in the output data message. This flag is detected in the control equipment which generates a warning message to the operator that the tag requires replacement.

Passive tags do not have a volatile memory and in consequence are frequently programmed at the time of manufacture. This is sometimes carried out by blowing predefined fusible links within the microchip. Alternatively an arrangement which makes or breaks links external to the chip is sometimes used. While the unit cost of passive tags is potentially less than their active cousins they suffer from the inflexibility that once programmed, their memory is fixed. There is also a hidden cost of programming and logistical control which should not be overlooked. At some point in the future the use of EEPROMs within passive tags may offer some interesting possibilities. For the present, however, this technology is unsuitable.

An interesting approach used by a few manufacturers has been the adoption of a quasi active system. Instead of a lithium cell, a capacitor with very low leakage is used to sustain a volatile memory. During interrogation the power to operate the chip and drive the output signal is derived from the interrogation pulse. Simultaneously the capacitor is recharged. These systems have the benefit of offering the customer a tag with a reprogrammable memory but without a battery. They are nevertheless subject to the same range and noise limitations as passive tags in operational use.

In addition to the custom chip and power source (if used) a tag requires its own input and output antennae. These take the form of a coil which may be either air or ferrite cored depending on the individual design. Passive tags tend to use large air coils for their input field antennae. This provides the most effective means of converting the interrogation field into useful internal power. Air coils can be wound with considerable accuracy by automatic machines. They have low values of Q and can therefore be used directly in production without the need for tuning. However, they are readily detuned in the presence of metal so may not be suitable in all applications. Ferrite coils are used in a great number of tags and have the merit of being very compact. They can achieve high values of Q while being relatively immune to the effects of nearby metal objects. However, due to the wide variations in permeability between production batches, ferrite coils invariably have to be physically tuned during assembly.

The chip, battery (if used) and other miscellaneous components will be mounted on a PCA. Depending on the individual design the antenna may either be positioned on the PCA or located directly in the case. Positioning of the antennae requires some care on the part of the designer. Failure to do so may lead to unacceptable levels of interaction which could seriously degrade the performance.

The completed PCA will be housed in a custom moulding usually made from ABS or polycarbonate. Before sealing, and depending on the application, the circuit may be encapsulated. This is particularly desirable in industrial applications where resistance to fluids and vibration is essential.

## 4.6 ANTENNA CONSIDERATIONS

The design of a successful family of read and read/write heads is critical to the operation of a tag system. Since the uses to which the product may be applied are very varied, it is very unlikely that any one design will suit all applications. For example, at one extreme a customer may wish to interrogate a tag at a very short range within a very confined space. Conversely another customer might want to read a fast moving tag at the maximum possible distance from the reading head. The two requirements would be met by separate antenna designs that were physically quite different.

Fortunately there are a number of common factors which simplify antenna specification. Firstly there is the question of range. Contrary perhaps to one's first thoughts practically all identification tasks can be performed at ranges of less than two metres. The reason for this is simple. The real requirement is to identify the person or object to which the tag is attached. Identification is generally associated with a need to carry out some form of control or process. This implies a physical action between main equipment at the interrogation point and the item to which the tag is attached. There will therefore be, in general, a system requirement to keep the distance between the main equipment and item as short as possible.

As a corollary there are two further points. In practically all instances a tagged item will move into an interrogation zone, be processed and then leave the zone. Secondly, since we are looking at a control process, items must of necessity be handled serially. There are very few examples of a control process which could be carried out effectively by one plant on two or more objects simultaneously. This would imply an ability by the identification process to determine which tag was attached to each item. It would additionally have to know that every item did in fact carry a tag and that all the tags present had been correctly accounted for. This represents a daunting task and for all practical purposes is unnecessary. It is a considerable simplifying factor therefore to limit the identification process to the presence of a single tag in the zone at any one time.

The second factor to consider is the orientation of the tag with respect to the read or read/write head. Fortunately in a great many applications this may be defined in advance. In such circumstances it is easy to ensure that tag orientation is optimised. However, in some applications the orientation of the tag may be entirely random. This might be the case, for example, with the identification of cattle where the cow is not always fully cooperative. In such circumstances the designer will frequently wish to make use of the effect of field curvature as part of his solution. This is illustrated in Figure 4.6.

The diagram shows the magnetic field from a coil energised from a low frequency source. Viewed from the side of the coil, the field pattern is coaxial at the coil centre but rotates through 90° at the coil circumference. Consider now the tag input coil A as it passes across the surface of the large coil. As coil A approaches the circumference it will be coaxial with the transmitted field from the large coil. This will induce a signal in coil A which will activate the tag. As the tag moves to the centre of the large coil the field will become orthogonal to coil A. At the centre no signal will be induced and the tag will revert to its quiescent state. As coil A moves out of the field it will again experience a coaxial field and reactivate the tag. Consider by comparison the effect on coil B. Tag activation will not occur at the circumference but will take place at the centre of the field. While this arrangement can cope effectively with any orientation of input coil in the x−y planes, it does not cope with the possibility of an input coil in the z plane. If a particular application should require reliable operation in all three planes a second large coil must be used. The second

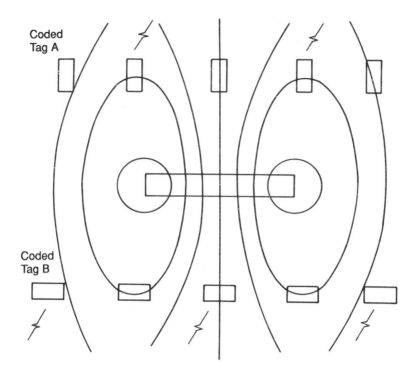

Coded
Tag A

Coded
Tag B

**Fig. 4.6** Activation of a coded tag passing across a reading head.

coil must be employed in such a way that its field does not interact with the first field. This point is important since simple interaction will produce the vector sum of the fields and the benefit of two coils will be lost. The two coils must therefore either be sited sufficiently far apart that the fields do not interfere. Alternatively the coils may be excited alternately by the use of a poling routine in the main control.

The third factor to influence antenna design is the time available to perform the interrogation. This is determined by the amount of data transfer which must take place and the signalling speed of the tag. Data transfer may involve simply reading a portion of data from tag memory. At a more complex level it might be necessary first to read a large quantity of tag data and subsequently to programme part of the tag memory with new data. The time taken to perform these two levels of data transfer will be significantly different and will vary from one tag system to another. Fortunately where large quantities of data transfer are required, it is often associated with a process where the tag remains stationary. Not infrequently, however, reading a small amount of data must be performed on the fly.

In order to read a moving tag it is essential that it should continue to be activated throughout the interrogation process. This means that the length of the interrogation field must be checked to ensure it is adequate for each application. The length required may be easily calculated from a knowledge of the time taken to transfer the data and the maximum speed of the tag. It is of course important to ensure that the direction and magnitude of the interrogation field remains broadly constant along the path where the tag will be read. The installer also should verify that the minimum distance between successive tags is such that no more than one tag will ever be in the field at any one time.

Inevitably there will be situations where a company's standard range of antenna is unsuitable for the application. In such situations it is possible for the installer to fabricate his own antenna system. Using conventional single core cable the antenna can be laid to the required length in the form of a loop. This is then tuned to the required resonant frequency by selection of appropriate capacitor values. The tuned loop may then be connected to the control unit in accordance with the manufacturer's recommendations.

Up to now we have only discussed the interrogation signal from the reading head to the input coil of the tag. Every bit as important is the return signal from the tag comprising the data. Fortunately the general law of reciprocity applies, which is that a good transmit aerial is a good receive aerial and vice versa. Provided, therefore, that the same general principles are observed for the return path as for the interrogate signal, satisfactory system performance should be achieved.

One final aspect yet to be covered is the programme or write signal. Extra care must be taken here since the programme mode is the most complex of the functions performed and any error could have serious system consequences. In general the maximum programme range should be restricted to 500 mm or less. This is usually sufficient to guarantee that a second tag is not simultaneously within range and inadvertently programmed. It is important also to realise that during a programme cycle the interrogation, read and write paths are all used. Programming is therefore a longer operation than a straightforward read cycle due to the need to verify that the data has been correctly entered. Apart from these additional points, the same general guidelines apply to the design of programme antennae as for reading heads.


## 4.7   CONTROL EQUIPMENT

The control equipment is the engine of any tag system. Broadly speaking controls fall into one of two categories. These are the interrogator or read only unit, and secondly the programmer or read/write unit. There are nevertheless a number of features which are common to both categories.

Firstly, reliable and consistent system performance is essential. Customers will automatically expect that the equipment will seldom, if ever, fail to read or programme a tag. They will also require that the probability of an incorrect read or write is infinitely small. To ensure this is the case, the designer must incorporate adequate error detection systems. Several techniques are available. For example, the data in the tag may be arranged to include some additional parity bits or checksums. Alternatively, use may be made of Hamming codes. Another method is repeatedly to read the tag data over a number of cycles and look for a match between cycles. A third approach is to include some redundancy, like, for instance, Manchester coding, in the transmitted data stream to or from the tag. In practice, in order to meet adequate system reliability, controls are likely to include a combination of these techniques.

The problem of system integrity becomes harder when considering the conditions which might exist at the extremity of tag range. In such a situation there is a real danger of a tag transmitting corrupt data due to intermittent operation. The solution is to incorporate some form of system hysteresis. This may be done in one of the two possible ways described below. In both cases the tag output signal is made sufficiently large that the receive path is not a limiting factor. This reduces the problem to one variable.

The first method is to incorporate within the design of the control, a means to vary the strength of the transmit signal. It is then possible for example to transmit interrogation pulses at different output levels. If the first pulse at low level produces an intermittent tag output, it will not match with a second pulse at high level. An alternative method which achieves the same result is to incorporate variable sensitivity in the tag input circuitry. In its quiescent state the tag will be in its low sensitivity condition. As soon as the tag is activated the sensitivity makes a step increase, thus ensuring a solid response during interrogation.

Another potential difficulty is to ensure that the system can cope with the ambiguity of more than one tag in the field at a time. Here, depending on the type of modulation and coding format, the designer must devise an algorithm within the control which discriminates between a single and multitag response. One very effective way of achieving this is to use a Manchester code format for the data output. The presence of two or more dissimilar identities is then immediately apparent.

In addition to the problem of system integrity there is the question of defining a suitable interface between the control equipment and its host. A number of standards are widely used within the computer industry which define hardware and format. Whenever possible these should be adopted since they will simplify integration. However, as yet, the RF identification industry has not developed to a point where there is any standardisation of signalling protocols or commands. The designer is

**Fig. 4.7** Access control using a coded tag system.

therefore left with no choice but to create his own. At some point in the future, standardisation must occur which will make interchangeability of tag systems a reality.

The data interface may be either serial or parallel. The latter is restricted almost entirely to the production and material handling markets. Where a serial interface is used the choice will be usually RS232 or RS422. There are also signs that the more recent standard RS485 might grow in popularity. While RS232 is probably most widely used, the benefits of RS422 with its multidrop capability give it some distinct advantages. Finally there may be a number of markets which have developed their own specific interface. This is exemplified by the access control industry which uses an interface standard based around the access card technology (Figure 4.7).

Turning now specifically to the interrogator, there are many situations where the user requires only to read the code in a tag. As a consequence many tag companies offer a range of read only units (see Table 4.1). These interrogators are less complex than programmers and are therefore less costly. Secondly since the read operation is much faster than programming, there is an opportunity in some applications to poll a number of reading heads from a single interrogator. This reduces further the cost per interrogation zone. Of course not all reading heads will be located necessarily in the same area. The interrogators must be capable therefore of driving reading heads over adequate lengths of feeder cable.

The single zone reader is a particular type of interrogator which is useful in certain situations. Due to its relative simplicity it can be manufactured very cheaply and is simple to operate. The device can be designed either for fixed installations or alternatively built as a hand-held reader for portable use. For permanent installations the single zone reader is cost-effective where either the total number of reading zones is small or the read stations are positioned at significant distances from the host. It is also a good technical solution where a large number of tags are frequently passing many reading points in a random manner. Under such conditions the polling routine of the multizone interrogator may be unable to cope.

The essential function of the interrogator is to read and validate the data in a tag and output this data to a host. There are, however, advantages in many situations of incorporating a limited amount of intelligence within the interrogator. This may be used, for example, to activate different outputs in accordance with predefined data held in the tags. Such a feature must include a capability for the user to modify the intelligence in the interrogator in accordance with his own changing needs.

The programmer unit must be capable of use in both an off line and on line mode. There are many situations where the user, for example, may wish to preprogramme tags prior to issuing them for use. To meet such a requirement a programmer should include its own keyboard, display and programming bed. Equally there will be other occasions where programming will be under central control from a host. This might further involve the need for a programming head which can operate remotely from the programmer. Due to the vital importance of correctly carrying out a programme cycle, external inputs and outputs to the programmer should be incorporated. These may be used, for instance, to initiate a programme cycle when a tag has entered the programming zone, and to move the tag out of the zone when the programme cycle has been completed correctly.

The format of the programme cycle must ensure failsafe operation. To minimise the possibility of any error a number of essential steps must be included. The first is to read the tag and confirm its presence. This step may also check that the ambient noise level is within acceptable limits and that only one tag is present in the field. Where a security code has been used within the tag database, the code may also be checked to ensure that the tag is authorised for use with that programmer. Additionally if the tag contains a battery its condition may be checked. Provided that the read cycle is valid the tag will be switched to its programme mode. Where there is no requirement for security this will be done by sending a simple command to the tag. However, where an objective is to prevent unauthorised tampering of the code, switching to the programme mode may require the transmission of a secure data pattern.

With the tag in its programme mode, the new data may be written to its memory. Depending on the size of the memory this may be carried out in one of two ways. If it is small it will be more convenient to rewrite the complete memory including its new data as a single operation. However, if the database is large, the memory will invariably be divided into addressable 8 bit bytes. In this situation it will be preferable to write to only those bytes which require change. With both techniques the write operation will be performed using the interrogation field from the programmer to generate the internal timing in the tag while a second signal will carry the data.

The final stage of the process is to check that the new data has been correctly entered into the tag memory. This is done by performing a second read cycle. The data received from the tag is compared within the programmer with that which was transmitted. Provided that a match of the data is achieved, the programmer will signal a successful operation.

## 4.8   APPLICATIONS FOR LF TAGS

By now it will be obvious that LF tag systems offer a very versatile and effective solution to a wide range of identification problems. In fact so enormous is their potential that there is only scope in this chapter to touch on their most obvious applications. Since also this is such a new technology it is unlikely that we have yet found the application where ultimately they will be most widely used. Nevertheless it is possible to define today a number of broad market areas where LF tag systems are already starting to make an impact. These may be grouped as follows:

• People
• Animals
• Vehicles
• Materials Handling
• Production.

Within each of these areas there is a further subset of applications. In order therefore to understand their breadth, we will look at each area in turn.

### 4.8.1   People

Perhaps the most obvious use for a tag system is as a means of access control. With a growing emphasis on security, more and more organisations are seeking to control movement of personnel within their premises. In the past this has been achieved by the use of keys, pin pads, swipe cards or personnel guards. These methods are acceptable where the user will only pass through a secure entrance once or twice a day. However, they rapidly become intolerable if the user is continuously passing through secure zones. This may be the case for example with a computer room, laboratory or accounts department. In such situations the benefit of a hands free tag system to the user in both speed and convenience is considerable. For the company the benefit is a more contented workforce and the knowledge that with such a user friendly device people will not be tempted to abuse the system.

Time and attendance is a logical extension of the access market for LF tags. Although this invites arguments about the 'big brother' syndrome, there are already a large number of swipe readers in use for this purpose. Tag systems, however, offer the advantage that they provide a faster throughput and are easier to operate. In addition, since the reading head may be buried within a wall it is impervious to dirt or wilful damage by an employee. This is not necessarily the case with a card reader which all too often is the unwilling victim of the chewing gum brigade.

LF tag systems offer opportunities for the use in personnel safety in a way which is not otherwise possible. For example, in areas of potential radioactive or medical contamination there is a frequent requirement to monitor and control access but avoid any physical contact. Similarly there may be situations where dangerous machinery may only be run in the presence of an authorised operator. A more mundane but equally vital area of concern is to control the unpredictable wandering of geriatric patients from nursing homes.

In all of these problems an LF tag system can provide a very cost-effective solution. The tag may be fabricated in the form of a credit card which makes it readily acceptable to the user. Depending on the precise requirement of the individual site, the reading head may operate either at short or long range. Short range (which covers distances of typically up to 50 cm) is most appropriate for access control and time and attendance where the need to control one person at a time is fundamental. Safety applications may call for a reading head which provides area coverage. Such a requirement is frequently best met by the use of a customised reading head in the form of an inductive loop.

Applications using LF tags and involving people do not generally have a major systems impact on a company's operation. For this reason they are relatively simple to sell, install and implement. Of all the potential market areas for LF tags this will undoubtedly be the first to use this technology in high volume.

### 4.8.2 Animals

A slightly surprising but nevertheless large market is that of animal tagging. Declining margins and fierce competition has forced today's farmer towards greater levels of efficiency. One area where improved efficiency is possible is in the automatic identification of dairy cattle. Each cow in the herd is fitted with a tag which may either be attached to its ear or fastened with a strap round the neck. Reading heads are fitted to automatic feed dispensers which are located both outside and in the milking parlour. A reading head may also be secured to a weighbridge at the entrance to the parlour. The reading heads are connected via the interrogator to the farm management computer system.

Knowing the identity of the cow at the feed dispenser it is now possible to provide exactly the right amount of food concentrate at correct intervals each day. The quantity of food delivered will be computed according to the weight of the cow, its milk yield and its point in the lactic cycle. The process therefore optimises the cost of feed against milk yield while simplifying the task of the herdsman.

A similar requirement exists for the identification of pigs. Here the need is more to ensure a controlled distribution of feed to the animals. This ensures that each pig receives its fair share and helps to reduce the incidence of fighting. Due to the difficulty of fixing a collar with any permanency around its neck, pigs are almost invariably fitted with ear tags.

### 4.8.3  Vehicles

The monitoring and identification of vehicles is an obvious application area for LF tag systems. However, this market presents its own peculiar set of problems. Vehicles are of random length and height. They may travel at varying speeds and frequently will not move along a closely defined direction of travel. In addition they contain their own internal vehicle electrics which may emit very high levels of electrical noise. Add to this the requirement to withstand continuous mechanical shock and extreme environment conditions and it will be apparent that this market presents some tough engineering problems.

Where it is necessary to identify a moving vehicle, it is preferable to position the reading head adjacent to a point where some physical control is to be exercised. For example this might be at a pay booth on a tollway or at the barrier into a factory or car park. By this means the vehicle's speed is naturally limited and if the road should narrow the position of the vehicle will more exactly be known.

An interesting question arises in such applications which is whether the real requirement is to tag the driver or the vehicle. In practice this can only be decided after a careful study of the particular application. If the decision is to tag the driver, there will be increased constraints on system flexibility. Inevitably the driver will be forced to slow down considerably while presenting his tag with one hand to an adjacent reading head. If instead the tag is mounted on the vehicle it may be possible to configure the reading head as a loop in the road. By making the loop as long as system constraints allow, the tag will be within the field for the longest possible period. This will enable the vehicle to be identified at the greatest possible speed. An added benefit of the buried loop is that it is extremely robust. Once installed it is almost indestructible.

The tag itself will in most cases be fitted to the front of each vehicle. This is an advantage since it makes it possible to relate identification with a known distance between the front of each vehicle and say a barrier. This would be less straightforward if tags were placed at the rear of vehicles of varying lengths.

The problem does not of course end at identification of the vehicle. In order to produce a working system the tag equipment must be interfaced with sophisticated control equipment. The control unit must operate either a barrier or lights for authorised vehicles and lower the barrier or reset the lights after each vehicle has passed. In addition it may have to offer alternative means of access like a coin machine for vehicles which are not fitted with an authorised tag. Lastly the control equipment should have a means of responding to any traffic violations.

Not all vehicle applications operate under quite such exacting requirements. The use of tag systems with automatic weighbridges is a good example. For some time weighbridges have been capable of recording the weight of vehicles on the fly. However, the ability to combine an LF tag system with a weighbridge means that both the vehicle weight and its identity can be correlated automatically by the control equipment. This simple marriage of two technologies produces a process which is quicker and more reliable than traditional methods as well as reducing manning levels.

Another useful application is the control and monitoring of fuel for vehicle fleets. Here a tag is securely mounted close to each vehicle filler cap. The fuel nozzles of the petrol pump at the company's fuel depot are fitted with intrinsically safe reading heads. When a vehicle arrives to refuel the driver inserts the nozzle into the vehicle filler which causes the tag to be read. If the tag data correlates with the information in the control system, the pump is activated and the driver is permitted to fill his vehicle. Depending on the particular design of the system the tag may be reprogrammed before the nozzle is removed. A fuel monitoring system of the type described can eliminate the considerable clerical effort in tracking vehicle fuel consumption and can greatly reduce pilfering of fuel.

### 4.8.4 Materials handling

The move towards more and more automation in materials handling continues. With it goes an increasing need to identify items automatically. There are of course many situations where materials handling problems can be solved readily using barcode technology. This is particularly so in areas where the position and orientation of each object is well defined and the environment is reasonably clean. However, there are many other materials handling applications which do not fall into this convenient category. These are the applications which frequently lend themselves to LF tags.

A good example of such an application is containers for fluids or gases. This can include beer barrels, containers for specialist chemicals and compressed gas cylinders. Frequently the cost of the containers for such materials forms a significant part of the total cost. It follows therefore that the manufacturers have a strong interest in wishing to track them. Additionally many of the contents of the containers will have a limited usable life. There is also therefore a benefit in being able easily to determine the expiry date. These requirements may be met readily with an LF tag system. The tag may be fixed on the outside of the container or permanently moulded within it. At the time each container is filled, its tag identity number will be read automatically. Simultaneously the expiry date of the contents may be programmed into the tag's memory. The container may then be tracked through the distribution process from the factory to the customer almost entirely free of human intervention. As each container is moved from one point to the next it can pass close to a reading head which will read its identity. This information can be passed automatically to a central computer. It is even possible to read a tag while it is rolling down a ramp as when being unloaded from a lorry. With such a system tracking becomes an automated process, with minimal errors and a substantial reduction in administrative costs.

The identification of tagged pallets and totes, which are used mainly to carry solid objects, is another application area (Figure 4.8). Pallets are frequently moved using fork lift trucks. These provide a convenient mounting point for a reading head and interrogator. If on line control is required it is possible to equip the fork lift truck with a radio which interfaces the interrogator to a host computer. With attention being focused more and more on the automatic guidance of fork lift trucks so the possibility of the automatic warehouse using tagged pallets becomes a reality.

While pallets are moved using fork lift trucks, tote bins are carried from one point to another by conveyor. Particularly in a large factory, conveyors can be very complex with numerous spurs and branches. The tote bins serve as a convenient device to route components round the plant. However, such a process is only feasible if it is possible to identify the tote at each junction and route it down the required path. The LF tag system is well equipped to perform such an identification role. A tag secured to each tote may be programmed with data which defines the routing instructions, the final destination of the tote, together with a

**Fig. 4.8** Pallet identification using a coded tag system.

description of the tote's contents. By this means it is possible to have a large population of totes all moving along the one conveyor and routed automatically to their own particular destination. While such ideas may sound fanciful they are fast becoming fact. Already trials on such systems are underway with plans for large scale complexes to go on line within the next few years.

Until now LF tag systems have made little impact on the materials handling industry. However, the long-term potential for their use in this market is enormous and covers a wide range of possibilities. This is illustrated by three specific application areas. These are the automatic routing of mail bags, the identification of container vehicles, and the automatic routing of passenger baggage at airports. All three problems potentially can be answered by LF tags. Only time will tell if this technology provides the most cost-effective solution.

### 4.8.5 Production control

Production control is probably the most challenging of all the potential market areas for LF tags. The need to minimise paperwork and to automate processes is present at all levels. The benefits are obvious − greater reliability, reduced manning and higher output. The risks are also clear. The introduction of a tag system which has not been adequately tested on a pilot plant could have a calamitous effect on a main production line. For this reason it is likely that production managers will introduce LF tag systems into their plants with caution. If, however, their performance lives up to present expectations then within the next ten years their use in industry will be universal.

The breadth of applications for LF tag systems within production is very wide. For convenience they may be grouped loosely under three main headings:

(1) Identification only.
(2) Identification and data.
(3) Portable database.

The first category is likely to be by far the biggest application area since identification of objects is fundamental to all production processes. Typical examples are the need to identify tool holders and product carriers. Both are simple straightforward requirements which exist in almost any major plant. Such items may be readily tagged and identified by reading heads sited at required monitoring points. Once an item has been identified, invariably some activity will be performed on it. The reading unit is likely therefore to be interfaced to a PLC which will initiate a sequence of operations.

This first category may use a non-programmable tag since an identity only is required which can remain fixed. At a rather more complex level there is the possibility with a programmable tag of including both identity and a limited amount of data. This data may be modified as the component moves through the production process. To illustrate the concept a line may manufacture fifty different product varieties from one basic component. At the start of the process each component would be fitted with a tag containing its identification number and product type. An area would also be allocated within the memory of the tag for status flags. As the component moved down the line its product type would be read at each work station, and input to the host PLC. This would control the cycle of operations relating to the product code number. On successful completion of the task, the PLC would instruct the read/write unit to set a predefined flag within the tag. At the end of the line the inspection station would confirm that all the flags relating to that product type had been correctly set.

Such an approach works well in straightforward applications which do not require a high degree of flexibility. However, in a process like the automobile industry there are more product variations than can be readily handled by product codes. Additionally the need to carry out unforeseen changes rapidly places a heavy reprogramming load on PLCs which in turn are controlled by a host computer. To overcome this problem the automobile industry are experimenting with the use of portable database systems. These use very large memory tags — typically up to 20 Kbytes which contain a complete set of vehicle build instructions. Because of the complexity of these instructions, each command will be loaded in ASCII in predefined addresses in the tag memory. At the same time it is necessary at each work station to programme the PLC so that it can respond to the full range of commands that it might receive via the read/write unit from each tag. This is likely to be a formidable task. For this reason the use of such devices will be limited to very complex manufacturing operations where the value of the tagged item is high.

Despite the wide spectrum of possible applications within production control, there are fortunately a number of parameters common to their general use. Firstly the position and orientation of each tag is known to a high level accuracy, and the tag will move along a predefined path. This generally makes it possible to operate over relatively short ranges (i.e. less than 15 cm) and optimise siting of read/write heads. Another simplifying factor is that the product generally moves along a production line at a slow speed, and may in fact stop at each work station while an operation is performed. Thus there is ample time in which to read and write to each tag. On the other hand there is the need to ensure that the system works with the utmost reliability. In the event that the tag control equipment should fail, automatic monitoring devices must alert the host immediately. Replacement of the defective component must then be possible with minimum delay.

Many manufacturers of LF systems for industry offer a family of products to cover as many sectors as possible. Tags are frequently available in a variety of shapes and sizes depending on: the environment in which they will be used; the available space for mounting; and the required read/write range. Similarly read heads and read/write heads will frequently be offered in a range of sizes and fitted with pluggable cable connectors for rapid maintenance. The control units will be designed for mounting within an industrial enclosure and again use quick fit connectors to minimise any down time. Since the control units will almost always interface to a PLC it may be fitted with two interface options. These most commonly will be RS232 for connection to one of the PLC serial ports or a parallel interface option for connection on the PLC to a bank of input/outputs.

## 4.9 CONCLUSION

Low frequency coded tag systems are a new and rapidly emerging industry. Currently their market is growing at a rate in excess of 40% per annum. They offer significant benefits as an identification and portable data device over other existing technologies in a very wide range of applications. They are already establishing themselves with success in the areas covered in this chapter. As they gain wider recognition they will become more and more part of our daily lives.

# Chapter 5

# Electronic Coins

## DAVID EGLISE

(British Satellite Broadcasting)

*IC − based 'coins' could be more useful than cards in some mass market applications.*

## 5.1 INTRODUCTION

Electronics has had a profound effect on the way we do things; it has altered existing practices and customs, it has even led us to do things we did not do before. The effects of what some have described as the third industrial revolution have been noticed in almost every sphere of life, from the workplace to the living room.

Those people who dislike technological change have always said the information revolution would not affect social traditions, but even they are starting to feel the wind of technological change with the gradual introduction of electronic funds transfer at point of sale, or eftpos, which is a system that allows a consumer to pay for purchases by directly debiting his bank account, transferring the money and crediting it to the vendor's account − all at the 'point of sale', be it a shop, garage or theatre box office.

Traditionalists have always said that this type of system could never replace hard currency because the cost of processing the sale makes it prohibitively expensive for small purchases, like buying a newspaper, paying the milkman or going to get your car washed.

But in the not too distant future microchips could replace the coins that wear holes in the pocket or weigh down the purse. Instead of a collection of change, the consumer of the next century will buy the daily newspaper with an electronic token − smart money (See Figure 5.1).

Unlike existing smart cards, which need to make physical contact with the interrogating mechanism to get access to the card's information, electronic tokens are derivatives of second generation smart cards − not only are they a unique shape but they require no physical contact to reveal their secrets. This gives these token devices several obvious advantages over their card-style ancestors. Currently smart cards need to make contact with the reading system and it must also be correctly oriented to be acceptable to the interrogator. This is usually achieved by incorporating

Fig. 5.1 Example of an electronic token.

electrical connections on the surface of the card to allow the reader to be linked electrically to the card's internal microprocessor. As a result, these contacts can sometimes be easily damaged or contaminated rendering the transfer of information impossible.

As a result, electronic tokens have certain benefits over current smart cards – for example, the requirement for contactless information transfer means the possibility of damage is drastically reduced and the device does not need to be inserted in any particular way. Because the electronic token does not normally require any electrical contacts on its exterior, the electronic devices within the token itself can be completely encapsulated within a ruggedised polymer material. This makes the device highly robust and durable and capable of operation in hostile environments as well as withstanding the curiosity of potential thieves. This method of packaging technology makes it relatively easy for the token to be offered in a variety of different shapes and form factors for different applications and geographic markets.

This can be realised in a situation where the system users in the USA would probably like to use an electronic token that is a comparable size to their most common high denomination, the 25 cent coin, while the United Kingdom would probably require a token with comparable dimensions to a 10 pence coin. By making the electronic token the same size as an indigenous coin, systems would not necessarily have to be mechanically altered.

What the electronic token can offer is the prospect of using a cashless payment system for low value transactions, such as buying a paper or making a telephone call at a public payphone − all without using conventional coins. Because conventional magnetic strip cards or first generation smart cards have an intrinsic high transaction cost overhead they are prohibitively expensive with transactions that involve very little hard currency.

Instead, the electronic token was originally conceived for use in public payphones as a cashless alternative. However this application placed several restraints on the system design. If electronic tokens are to be used in conjunction with public telephones they have to be extremely secure. One can imagine that there would be large numbers of tokens in circulation at any one time and the interrogators would be installed in remote locations. This situation makes them vulnerable to investigation by would-be *thieves*. These tokens must also be able to function over wide environmental limits and still maintain a high level of reliability. If tokens are to be used in public payphones throughout the country, then the devices must be able to operate from a telephone line supply, requiring the token and its interrogator to have a very low power consumption. Since the payphone development, many other applications have been discovered, which do not necessarily require these generic features, but they often offer significant benefits even if they are not absolutely vital.

## 5.2  BASIC SYSTEM REQUIREMENTS

To provide these unique features, the electronic technology within the token comprises a 'secure' 8 bit microcontroller, 2 to 8 Kbit of 'secure' electrically erasable memory and an interface communications circuit. Alternatively, the token can be used in different applications, where a microcontroller is not absolutely necessary. In this case the token may be configured to consist of a communications circuit for the power and data interfaces, simple logic and 'secure' electrically erasable memory.

The use of a 'secure' microcontroller gives the system added security because it allows the system designer to incorporate sophisticated cryptographic techniques to ensure that the integrity of the token and reader is at its highest possible level, but, at the same time, flexible enough to be altered to suit different applications. The alternative use of simple logic provides some level of security but does not offer the overall flexibility of a microprocessor-based token, so it is confined by design, to specific applications.

As mentioned previously the electronic token has no external contacts and certain applications demand the device must have an extremely low power consumption. To satisfy this requirement a novel and innovative means of power and data transmission had to be developed. The final

**Fig. 5.2** Token showing antenna coil.

embodiment involves both the interrogator and the electronic token incorporating a simple antenna coil. (Figure 5.2). This allows the devices to transmit power and communicate when the two are brought into close proximity with each other. One innovative part of this system is the circuitry within the token which allows it to communicate in a 'duplex' mode with the interrogator without the need for an internal power source within the token and still meet the power requirements of the overall system.

This has been achieved by an active 'zero' loss field effect transistor bridge connected across the antenna coil of the token. The bridge handles power and data signals derived from a 113 kHz carrier transmitted by the interrogator antenna. Data is effectively transmitted to the token using an amplitude modulation technique, i.e. by omitting certain half cycles in the carrier. This technique directly affects the internal power supply to the token circuitry, so when these omissions occur, a reservoir capacitor connected in parallel with the token's antenna coil, maintains the continuity of power to the internal circuitry. These omissions in the carrier are subsequently detected by a special amplitude modulation detector which reconstitutes the information and passes it to the token's 'secure' microprocessor.

When the token requires to transmit information back to the interrogator, the microcontroller in the token will operate a device that shorts out the 'zero' loss transistor bridge for a precise period of time. This event effectively short circuits the token's antenna coil and consequently causes the amplitude of the carrier at the interrogator to drop for a short period of time. The effect is caused by the absorption within the token being reflected through to the interrogation coil and creating subsequent power losses in the coil of the interrogator resulting in a predictable voltage change at the output stage. The interrogator has within the output stage, a sensitive carrier amplitude detector which detects the 'dips' in carrier amplitude; this specific modulation being interpreted as data. Consequently a quasi-synchronised sequence of 'dips' in carrier amplitude is used to communicate information from the token to the interrogator device.

## 5.3 APPLICATIONS OF ELECTRONIC TOKENS

The combination of a programmable microcontroller and electrically erasable memory gives the tokens great versatility, which allows them to be used in several different ways, as well as a multitude of applications. As a method of payment, the tokens can be used either in credit or debit mode. In credit mode, the token can store information such as a credit limit, personal identification number, or PIN, and other details of the token owner. When the token is used as part of a transaction, the communications protocol is such that the interrogator can request a PIN from the token's owner and validates it against the PIN stored in the token's secure memory. Consequently the PIN is not transferred beyond the transaction terminal and into the wider system environment. This technique gives the system an exceptionally high level of security at the point of transaction. This process also minimises infrastructure overheads because an interrogator does not have to refer back to a central database to check the PIN entered to a look up table. When the purchaser is authenticated, the transaction can be authorised by the control system by one of several methods. The actual method would probably depend on the specific application, but the most common ways of transferring the value of the transaction would be directly debiting through an eftpos system, in a similar way that current credit card schemes operate. Working in a credit mode tends to maximise business use and revenue but it has certain drawbacks. A credit system is difficult to operate in a country that has an unstable currency, and the system requires a sophisticated, centralised database.

Alternatively the token can be used in a debit mode. In this type of operation the electronic token would be accredited with some pre-programmed value. When it is used as a payment device to buy goods or services, the value of the purchase is deducted from the token's original value. The token, with its residual credit, is returned to the purchaser. When the token is devoid of value, it can be returned to the issuing authority. There it will be given more credits by a high security seeding process that is not utilised in the field and reintroduced into circulation. Tokens with a large enough memory can operate in both credit and debit modes, with the user choosing whether to charge a purchase to his bank account or to pay for it with the monetary value on the token. In debit mode, payment is immediate and can be easily audited by the authority controlling the system. It is not necessary to have an established credit card style system in a country to support a debit system because the debit tokens can simply be purchased from a point of sale. Credit mode systems, however, require a rather more sophisticated and costly operating infrastructure. But the hardware that is used in a debit mode system is usually more complex because it is designed to be able to alter the token's secure memory and, in most cases, extra levels of security have to be included to prevent system abuse.

Certain benefits of using electronic tokens can also be realised in areas other than just low value transactions − tokens can offer benefits over existing methods in areas such as data collection, identification and access control. For example, every time a token is used to make a telephone call, the token could store details of the call, such as duration, destination, cost, time of call and other information. When the token is returned to be revalued, the data could be extracted to give useful information on users' habits and possible trends, giving the benefit of providing information that could be an important business tool.

Electronic tokens could also be used to allow access to various events or programmes on television, or they could be used as access devices to open doors. In either of these cases, access could be selective, with some devices opening all doors, others opening just a few. In the same way a child could be prevented from watching certain television programmes or even all programmes after a specific time. The data stored in the token would inform the monitoring system whether or not the token holder is allowed into a certain room. In the television application, the interrogator system could determine from the data stored within the token whether the viewer is old enough, or whether they are eligible to see a particular program or film.

## 5.4  LOW VALUE TRANSACTIONS

However, the application area in which electronic tokens offer the most benefits, is that of low value transactions. The development of the necessary technology has coincided with the growing interest in alternatives to payment by cash, particularly in low value transactions. This interest has arisen because operators of vending systems in general — whether they are public telephones, drinks dispensers or ticket machines — require to maximise income by minimising system costs and overheads. Consequently, by implementing a system based on electronic tokens, a vending machine operator can significantly reduce theft/vandalism from the system. This is realised by the fact that a vending machine using an electronic payment method offers little incentive to thieves because it does not collect money. Another problem for the conventional vending machine system operator is the cost and effort of physically collecting the coins when they have been deposited in a machine. This is an expense that could be eliminated by using electronic methods of payment. For example, the public transport system in London collects several metric tonnes of coins per day. This is a typical application, where the use of electronic tokens could offer drastic savings to the cost of collecting money as well as providing several user benefits. One important advantage that electronic tokens have over conventional currency is that their physical size can be constant relative to their changing value. What this means is the selling price of goods or services provided by a machine, can be altered in such a way that the payment medium remains the same. An operator of a machine does not have to get involved in modifying the mechanics to accept additional/new coins. With electronic payment, the price setting within the vending machine can be altered simply to charge more for its goods or services.

One obvious advantage of electronic tokens over current magnetic or smart cards is the cost of the interrogator. A token reader is currently around 10% of the cost of a conventional magnetic card reader. Further cost advantages can be achieved by using tokens instead of other cashless payment methods because a token interrogator can be designed to be retrofitted to an existing coin operated machine — by default, card methodology cannot always be cost-effectively retrofitted to existing machines. The main reason is that it can incur major design changes to the existing mechanical features of machines.

As the electronic method of payment can be made the same size and shape as a coin, it can use the same slot, with only slight modification to the coin validation circuitry required. As a result a vending machine can accept both conventional metallic coins and electronic tokens. This benefits both the system operator and consumer. It allows the operator to install machines that can accept both forms of payment, but does not alienate consumers that choose not to use electronic tokens or vice versa.

Because the electronic token can be inserted into a machine in the same way as a conventional coin it can be placed in proximity with the token interrogator system without mechanical assistance. In cardbased systems, there is the requirement for a transport mechanism to physically move the card from the entry slot to the interrogating component. As a

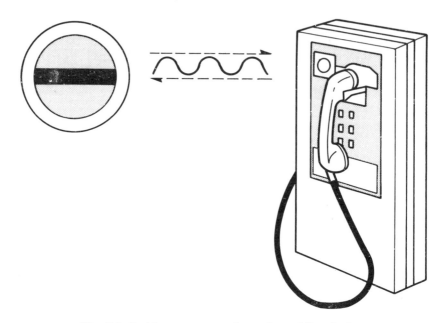

**Fig. 5.3** Cashless payment scheme for public telephones.

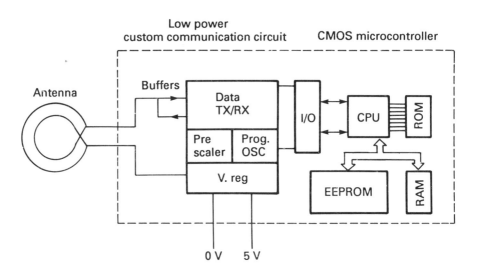

**Fig. 5.4** Functional block diagram of an intelligent token.

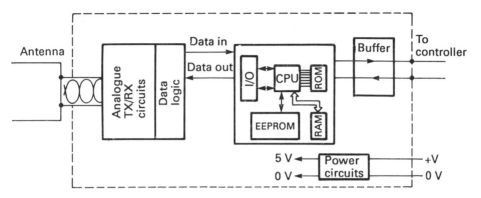

**Fig. 5.5** Functional block diagram of an interrogator.

consequence, a token interrogator has no prime moving parts − unlike its card equivalent − and is therefore intrinsically more reliable. Taking this a stage further electronic tokens can be applied without the use of a coin slot, that is to physically place the token on a defined area of the surface of the equipment case. Because the transmission of information between token and interrogator is via inductive coils, the two devices can be separated by a distance of up to 10 mm. This separation criterion can include certain materials positioned between the two, as long as they are not conductive. This configuration means that the interrogator can be positioned such that it would be impossible to reach by anyone externally. On the other hand if a conventional coin slot was used the interrogator would be equally difficult to get at by a potential vandal, but the slot could be 'blocked' rendering the machine useless. The absence of a slot would avoid this potential problem.

One of the main commercial issues with electronic tokens at present is that they have yet to achieve a level of cost that would facilitate widespread use. The present token design uses separate electronic microchips, but these circuits are expected to be integrated onto one piece of silicon in future versions of the electronic token as technology advances. Because the silicon circuits represent a considerable percentage of the total cost of the token, reducing the amount of silicon can substantially reduce the manufacturing cost of the tokens. This situation is dependent on the progress made in the areas of semiconductor design involving the integration of different technologies onto the same piece of silicon.

Electronic tokens can offer more benefits than just minimising the cost to the system operator of low value transactions, they can offer advantages to the system user. Electronic tokens give more services to the users and cut the system costs for the operators. Take, for example, the original application for which electronic tokens were intended, public telephones. In public payphones, cashless payment would allow telephone companies to increase the convenience of payphone usage. Callers would not need large amounts of change to make long distance calls and, depending on what type of electronic payment is used – credit or debit – calls from a payphone could be added to a domestic or business account. In this type of system the payphone can be perceived as an extension to a domestic or business telephone – an idea that is being pursued by a number of telephone authorities.

The advent of liberalisation in the payphone sector in many countries, such as the USA, is stimulating an improvement in telephone services and a need to offer something extra that differentiates one provider from another. Because of this and the high level of competition there is significant opportunity to introduce various methods of cashless payment. Tokens can offer an ideal introduction to cashless transactions using payphones. The necessary equipment to start the service is very small, because the token can use the same payphone as coinage, albeit with an additional internal component. In recent years, the installed life of payphones has dropped from ten to five years. This is due to new technology and increased vandalism which has accelerated obsolescence. Electronic token payphones can offer greater resistance to vandalism and, because of the flexibility of the system, they can easily be modified when changes in charges occur (Figure 5.6).

To give an idea of the potential market for electronic tokens in public telephones, there are at present around 2.75 million payphones in the USA, the UK, France and West Germany, and of these less than 100,000 are cashless payphones. However this is expected to rise to nearly 1 million by 1990. While the payphone operators are motivated to eliminate cash, the public are not only slow to accept the cashless alternative but the cost of present cashless systems is restrictive. Current solutions to eliminating cash fall into three main categories: small debit cards, swipe card reader credit card systems and key entry or operator-based account call systems.

The payphone operators believe a future solution should comprise of elements of all three systems as none of them are completely satisfactory on their own, the first two being expensive to operate and the third requiring the consumer to remember a 25 digit number to instigate a call. The inherent capabilities of the electronic token can more than adequately match the needs of a payphone operator. It allows credit and debit transactions without the need for expensive computer networks and also enables operators to gradually replace payphones that accept cash only with ones that can accept both or are cashless only.

Initial studies show that the cost of upgrading 100,000 cash payphones to accept electronic tokens and the cost of the tokens is 65% less than

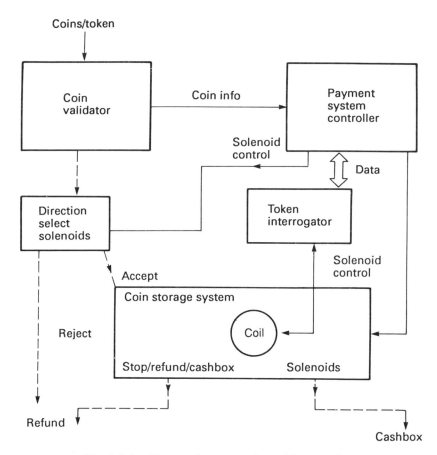

**Fig. 5.6** Intelligent token operation with a payphone.

implementing an equivalent system using the current 'debit' card system. If, instead of retrofitting the payphone, one was to introduce 100,000 new phones, then the electronic token payphones would be 23% more cost-effective than the debit card alternative. As the electronic token system does not necessarily require an established base of agency credit card users, it can exhibit considerable advantages in third world countries, which could lead to a significant growth potential.

Another low value transaction area which can reap similar benefits to those for payphones, would be the conventional vending machine sector. Like payphones, this sector is also motivated to reduce coin handling as it is expensive and labour intensive. In debit usage, tokens could offer immediate payment up front – a factor that machine operators consider attractive because it would improve overall profitability and could bring cost reductions. Cashless payment under certain conditions can also sig-

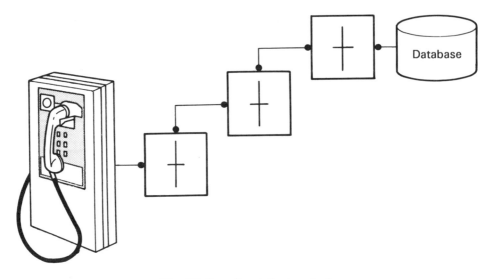

**Fig. 5.7** Security at the terminal.

nificantly increase sales by promoting 'impulse' purchases to take place by reducing some of the inhibitive processes, such as not having the correct coins or having insufficient change. As with payphones, the use of electronic tokens in vending machines can also eliminate the effects of changes in coinage and the price of the goods sold through the vending machine. These not only force operators to update expensive coin validation systems, but also involve additional inventory and technical support costs.

Electronic tokens as a method of payment for goods and services from vending machines, would probably not be suitable for those machines sited in public places, because of difficulties in issuing and controlling tokens for passing trade. However, most of the 4.5 million food, drink and cigarette vending machines in the USA and Europe are within controlled sites, such as factories or office blocks. In these examples electronic tokens could offer operators similar benefits to those that could be achieved by payphone operators. The overall potential is not as great as for vending machines, but nevertheless offers considerable potential.

The photocopier market can also utilise electronic tokens but in a slightly different way to payphones and vending machines. In general, photocopiers produce high revenues in three main areas; copier hardware, media for copies and technical services. The majority of copiers in service are leased, with leasing payments being calculated on per copy basis. The photocopier machine operators are generally responsible for the meter reading indicating the amount of copies produced but they have little motivation to record and report readings either promptly or accurately.

Electronic tokens offer the photocopier leasing companies a solution to this conflict of interests. Tokens can be designed to be programmed securely with information related to a number of copy credits. These programmed devices can be despatched to the machine operator on site, who downloads the tokens into the photocopier control system, which effectively stores this information as copy credits. Every time the copier is used, the number of copy credits is decremented. Eventually the credit in the copier machine is exhausted and the machine will cease to operate. The incentive is to maintain the continuity of copying service, so the operator now uses tokens as a data carrier and is forced to apply to the issuing authority to purchase others in advance. Electronic tokens can also offer extra features like monitoring the amount of copying done by certain individuals or departments – either in a leased system or a company's own system. Another use of tokens in the photocopier market is as payment for copier services in controlled areas such as universities or libraries – primarily a low value transaction like payphones or vending machines. Although the overall potential of electronic tokens in the photocopier market is restricted to the medium to high volume copying machines, there is still a degree of synergy comparable to that for conventional vending machines because of the financial advantages tokens can offer.

Although there is an enormous potential within the three applications already mentioned, the list of applications for electronic tokens does not end there. Several other areas, some of which have already been touched upon are already being considered. Because the token can store information securely and easily, it can be used as an 'access key', controlling access in a variety of different situations. One can consider electronic tokens being used to gain access to cars and households, hotel rooms and offices or factories. The token's programmability and flexibility, can be used most effectively in these types of applications, particularly in areas where individuals need to be authorised or even monitored in very specific ways i.e. it can be configurative. The token can also include identification data, so that when an employee or hotel visitor wants to gain access to a door, they must authenticate who they are by giving a personal identification number that matches the one stored in a token before authorisation is given.

In a hotel environment, as well as making sure a visitor only enters rooms to which he/she has legitimate access, a secure access device could be used also as a billing medium. This implies that when a hotel guest requires a drink at the bar, he uses his room key to log the transaction; when he wants to watch a film on television in his room he uses his key to log the event. This offers the hotel and the guest the capability of monitoring the activities of guests and charging them accurately for what they use. In addition, the visitor carries a running total of his expenditure

**Fig. 5.8** Internal view of an electronic token.

in the token, he does not have to wait while the staff calculate the bill when he departs the hotel, including any last minute expenditures.

At a place of work, electronic token devices can also provide an adequate level of security to offset industrial espionage and make sure that the only people that gain access to certain parts of a building are those who are supposed to. By merging an electronic token system with an employee identification scheme, such as signature verification, a very secure environment can be achieved.

Like hotels and offices or factories, leisure centres and sports clubs can all benefit from using electronic tokens as an enabling device. Instead of a membership card, an electronic token can give access to the facilities of the centre or club without the need for a person to monitor movements into different facilities. The intelligence of these token devices can be utilised in several ways such as booking of facilities whilst providing a useful management tool to monitor the usage of the leisure centre or sports club by various sectors of the community automatically.

The areas mentioned in which electronic tokens can be used are just the tip of the iceberg. As well as payphones, vending machines, photocopiers and access control, the electronic token is being considered for use in inventory control, gas and electricity metering and even access to satellite television programmes. As the awareness of the capabilities of electronic tokens grows, many more areas in which these devices can be used will be thought of. There is no doubt that this type of technology may become part of our everyday lives because it offers potential operators so much for comparatively little cost, particularly in the area of low value transactions.

## 5.5   SYSTEM CONSIDERATIONS

There are many reasons why electronic tokens might be used in the near future. Tokens offer an adequate level of security because of the device's capability of processing cryptographic information, so that if an unauthorised person managed to get into the system, the information would be illegible. A further benefit of storing information within the device is that the interrogator does not always have to refer back to a central database, it can validate certain levels of information itself remotely – which reduces the processing overheads associated with a transaction.

The lack of physical contacts increases the reliability of such a device, and therefore the system. It means tokens can be produced in almost any shape or size required (Figure 5.9). Making tokens the same size as coins has obvious advantages because there is then no need to modify coin slots, and only slight alterations are needed to the coin validation system. Therefore a payphone that can handle cash and cashless transactions without great expense to the operator, can be effectively realised.

Obviously the previous advantages have relevance to a cost advantage – better security means less system compromises – but contactless tokens are inherently more cost-effective than cash or smart cards as a means of payment. Like first generation cards, tokens eliminate cash collection costs, but unlike cards, 'intelligent' token systems can offer

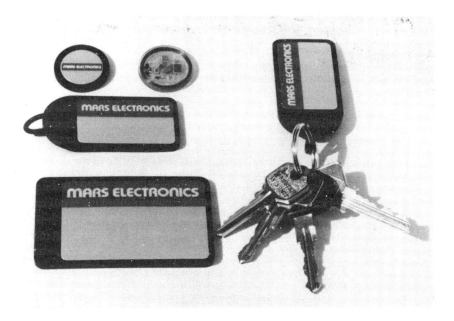

**Fig. 5.9** Different form factor of electronic tokens.

cheaper system costs and running costs. Even compared with low cost disposable debit cards that use magnetic strip technology, the contactless intelligent token is as cost-effective because it can be reprogrammed. So although the throw away card costs less to produce, it can only be used once, whereas the token can be reprogrammed and used many times. Cashless methods of payment also increase sales by eliminating some of the barriers which prevent purchases, such as not having the appropriate coins or the machine being too full of change to accept any more.

What makes electronic tokens ideal as a means for buying goods or services that are usually a low transaction, is the fact that a system based on these devices does not require massive processing capabilities and a complex infrastructure in which to operate. As a result, the cost of processing a transaction is extremely small, making it possible for cashless payment systems to enter the realm of the low value purchase – something that has, until now, been prohibitively expensive. The fact of the matter is that business operators of a variety of systems, from telephones to satellite television, all require a transaction system that reduces costs and offers real benefits to their users.

The traditionalists will soon have to start coming to terms with using electronic tokens to buy their daily newspaper because before very long these devices could become commonplace. They may even replace conventional coins altogether!

# Chapter 6

# Secure Transactions with an Intelligent Token

W. L. PRICE AND BERNARD J. CHORLEY

(Head, Data Security Group, National Physical Laboratory)

*The intelligent token is a type of supersmart card providing security for electronic banking and commerce.*

## 6.1 INTRODUCTION

The NPL intelligent token was conceived and developed as part of a research programme, beginning in 1982 and continuing for eight years, sponsored by the Tokens and Transactions Control Consortium (TTCC); the latter was set up by the British Technology Group and the UK Department of Trade and Industry. The research programme was carried out at the UK National Physical Laboratory and the members of the consortium included organisations from amongst the suppliers and users of access control and data security technology. Descriptions of the NPL intelligent token can be found in conference papers by Price and Chorley [1,2]. Studies of a similar concept have been carried out by the OSIS organisation (now TeleTrusT) [3], though, as far we know, this system has been demonstrated only by simulation

We give here an account of the general design philosophy of the NPL token, together with some details of the design itself and a summary account of the potential applications.

In the context of access control (to computers and, indeed, many other systems) there is usually a need for a reliable means of personal identity verification. In many computer-based transaction processing systems there is almost always a further need to ensure the integrity of transaction messages initiated by or on behalf of the system user. The design philosophy of the NPL token aims to meet these two needs in the one device.

It is becoming increasingly common for access to computer systems to depend on knowledge of some password, often coupled with possession of a token; a very common experience is use of a token in the form of a bank plastic card, associated with a password in the form of a personal identification number or PIN. Unfortunately the security of the plastic card, with magnetic stripe, leaves a great deal to be desired; it is far too easily copied, though anti-forgery features are added by the manufacturers. The known insecurity of the magnetic stripe card is one of the reasons behind the development of the 'smart card', which is the subject of much of the present volume. The smart card has the advantages of far greater storage capacity, coupled, in some cases, with internal processing power. Smart cards are used in many applications including payment for services, payment for goods and operation of personal or corporate bank accounts. A consensus view holds that smart cards are far less easily copied or forged than magnetic stripe cards.

Of course, if a smart card designed to be used without password is lost, then anybody finding it can use it; this is the case with 'smart' cards designed for telephone prepayment applications. Therefore, in applications requiring higher levels of security, such as in accessing bank accounts, a PIN is usually associated with the card. Correct response from the card depends on correct presentation of the PIN to the smart card terminal.

It is well known that the security of systems which depend on PINs leaves a lot to be desired. Habits of people with PINs can be very bad, such as writing them on the cards or sharing them with other people; it is not uncommon to wait at a bank ATM whilst the person in front presents a number of cards on behalf of colleagues, with the relevant PIN for each card. The alternative to the PIN is personal identity verification by one of several possible biometric techniques. Biometric methods measure the way in which the card holder carries out a specified task, such as signing one's name, or else measure a physical characteristic of the card holder, such as fingerprint pattern. The aim of the biometric system is to make impersonation extremely difficult. Several problems hinder the introduction of biometric methods; these include variability of results, relatively slow response in some cases, unpopularity with the users, etc. Until more acceptable biometric methods are available, we fall back on the PIN, which is at least better than using a card with no password.

Theft of PINs by thieves must be prevented at all costs. Such theft can take place in many ways, including bugging a transaction terminal to collect PINs. Whilst the risk of a bank automatic teller machine being bugged for this or any other purpose is very small, this cannot be said for the kind of terminal which we are beginning to find on retail counters in shops. Cost considerations dictate that the degree of tamper resistance that can be built into retail terminals is low. PINs presented on low security retail terminals may therefore be subject to disclosure.

If a PIN associated with a magnetic stripe card is stolen, then the thief may be able to create a false card and thus deceive the access control system. If a PIN associated with a smart card is stolen, then it is likely that the thief can only profit by his action if the card is also stolen. This must be regarded as a real possibility and therefore there may be an advantage in a system which gives better protection to the PIN.


## 6.2 DESIGN PRINCIPLES OF THE TOKEN

One way in which the PIN can be given better protection is to provide a keyboard for its entry which is under direct user control. Thus the PIN is not entered on the keyboard of the retail terminal, but upon a keyboard specially mounted on the token itself. The first design principle of the NPL intelligent token is therefore that the PIN should be presented to the system on a keyboard integral with the token. We shall later describe how the token is able to check the PIN validity and then satisfy the terminal that the PIN was valid without actually disclosing the PIN to the terminal. Clearly it is not sufficient for the token to receive the PIN on its keyboard, check it and then send a message saying 'the PIN was correct' to the terminal, since this message could be sent by a false token with an incorrect PIN. The message must be such that the terminal can rely upon it.

In a transaction processing system the user is usually dependent on displayed information on the terminal; the question is whether the user can always trust this information. It is a common experience to motorists to buy petrol from a pump with digital light emitting diode display; elements of these displays frequently fail and an amount (petrol or value) is displayed which is different from the correct value. It is not inconceivable that a retail terminal display might be deliberately altered by someone seeking to defraud the system owners or users. Again this brings us back to the customer's personal token. If it were possible to display the transaction details, particularly the amount of money to be committed, on a display under customer control, then this problem would be much reduced, at least from the customer's point of view; comparison of token and terminal displays may give added confidence. For this reason the NPL intelligent token carries its own display for communication with the customer, which is the second important design principle of the token.

We have indicated earlier the importance that must attach to the integrity of transaction messages which control the movement of funds between user and retailer accounts. It is frequent practice to protect transaction messages by encipherment techniques, often by authentication based on symmetric encipherment algorithms. An alternative and attractive method of ensuring integrity is based on the digital signature derived from an application of public key cryptography. As we shall see, the NPL token is able to satisfy a terminal that a correct PIN has been offered without disclosure of that PIN; this property depends upon the application of a digital signature. The basic token design therefore includes the ability to calculate digital signatures using a stored secret key. Since the ability to calculate signatures is a fundamental requirement in the device, it is convenient to apply this ability to calculating signatures on transaction messages authorised by the token holder. Transaction messages signed in this way can be checked anywhere in the transaction processing system where a reliable copy of the corresponding public key is available.

## 6.3 REALISATION OF THE TOKEN DESIGN PRINCIPLES

We have now identified the three fundamental design principles of the NPL intelligent token – integral keyboard, integral display and ability to calculate digital signatures. We proceed to discuss the ways in which these design principles have come to be implemented in physical and logical terms. The central part of the design is an implementation of the RSA public key cryptosystem [4]; this software implementation runs on a fast signal processing chip, the Texas Instruments TMS32010.

Personal identity verification (more strictly PIN verification) begins with presentation of the token to a terminal by the user; the terminal senses the presence of the token and generates a random number which it sends as a challenge to the token; at the same time the token signals to the user (using its own display) that the PIN must be input on the token keyboard. The token is designed to check the PIN and, if the PIN is correct, to sign the random number just received from the terminal using, for this purpose, the secret RSA key contained within the token. (Should the PIN be incorrect, the signature process does not take place and the user is given a limited number of attempts to get the PIN correct, failing which the token is disabled.) Having produced a transformation of the random number by the signature process, the token returns the transformed number to the terminal. The terminal, after having generated the initial random number challenge and while the token is preparing its signed reply, will have sought the public key corresponding to the token; this can come either from a reference source of public keys or can be supplied by the token itself in the first exchange of data with the terminal, in which case the version of the public key is supplied already signed by the secret key of a superior authority (the public key of the authority must then be known to the terminal). Given the public key corresponding to the token, the terminal can check the validity of the returned signature of the random number challenge; correct signature implies correct PIN presentation on the token keyboard. Figure 6.1 illustrates the sequence of events in identity verification.

Because the token is capable of generating RSA signatures, it is a simple extension of its functionality to permit the signature of transaction messages. In a retail point-of-sale system these messages would be pre-

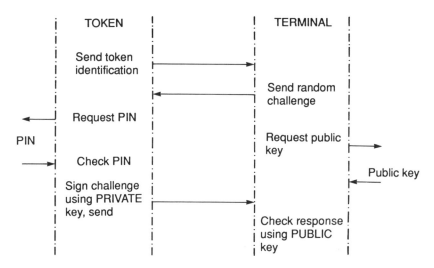

**Fig. 6.1** PIN checking, token challenge and response.

pared on the retailer terminal and sent to the token for approval by the token holder (inspection in the token window by the user) and, if approved, signed by the token and returned to the terminal. The terminal can check the validity of the signature and then send the signed message to a transaction processing centre. The signature validity can be checked by any entity in the system having access to the public key corresponding to the signature token. To avoid replays of transactions, it is necessary to include a time and date field in the message. Transaction numbering does not lend itself conveniently to prevention of replay for token originated transactions; tokens accessing multiple services would require a serial number for each and hosts offering services would need a number for each token in valid issue. Figure 6.2 illustrates the sequence of events in transaction signature.

It is an interesting extension of the design that the initial random number challenge may be omitted and replaced by the transaction message. In this case the protocol is shortened by arranging that the identity verification is checked by the signature on the transaction message.

The ability of the token to sign messages can be extended to cover messages in general; the application of the token is not restricted to value transactions.

## 6.4 THE PROTOTYPE TOKEN

The NPL intelligent token was created in prototype form in a unit measuring 36 cm × 15 cm × 2 cm; this device contained 21 discrete

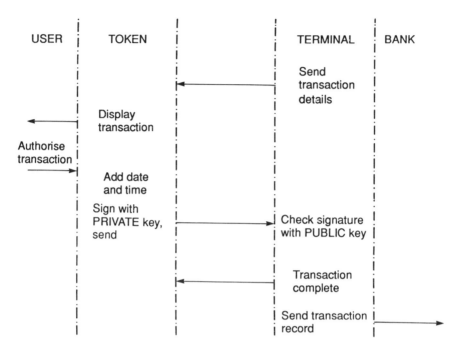

**Fig. 6.2** Transaction authorisation and signature.

integrated circuits, including the TMS32010 and an Intel 8085 to act as controller. Battery maintained RAM was provided for storage of parameters such as keys and a record of transactions.

Consideration was given to a semi-custom designed RSA processor chip in the early days of the project. It was considered at that time that the technology was not readily available for such a device and so an alternative method of implementing the algorithm was sought. Texas Instruments had just released the first in a series of fast processor chips designed mainly for signal processing applications. This device, the TMS32010, is a 16-bit microprocessor with an instruction execution time of 200 ns. The instruction set includes a signed 16-bit multiply executed in one cycle. Although not ideal (overflow and the sign bit caused problems) this processor was programmed to perform the RSA calculations. The new design continues to use this processor.

The TMS32010 has limited program and data memory spaces of 4K words and 144 words respectively. Further, there is no high-level language compiler and the speed of the processor makes it difficult to interface to slow peripheral chips. For these reasons, all the remaining functions of the token were placed under the control of a second processor. This split has several advantages; the two processors can work in parallel, thus reducing or even hiding the RSA calculation time, a high-level language can be used for the application software, none of the TMS32010 memory resources are wasted, and peripheral interfacing is easier and uses fewer components.

Creation of the prototype enabled the development team to demonstrate the correct functioning of the device in applications such as access control, point of sale transactions and signature of alphanumeric messages. Since the prototype was comparatively large, the next stage was to engineer a smaller version. In order to reduce the size, the functions of a number of the separate integrated circuits were absorbed into one application specific integrated circuit (ASIC). The chip count was thereby reduced to 11; further space was saved by using surface mounting technology. The result, produced in collaboration with Texas Instruments, was a device similar in size to a medium sized pocket calculator (about 14 cm $\times$ 9 cm $\times$ 1 cm).

In the new version, the Intel 8085 is replaced by a member of the TMS7000 series of 8-bit microprocessors. As before, this processor also controls the RSA processor and all peripherals, maintains the secret data and runs the application program. 32 K bytes of program memory are available, most applications to date have used only half of this amount. 8K bytes of battery backed RAM are built in for the storage of keys and other data needed for applications.

All of the decoding logic, address latching and bus de-multiplexing for both processors have been reduced to a single semicustom chip designed at the NPL and fabricated in 1.8 micron CMOS gate array technology by Texas Instruments. Figure 6.3 is a block diagram showing the important physical features of the token. The display is a 16 character by 2 line liquid crystal display and the keypad consists of 4 rows of 3 buttons. The reasons for including these on the token are given elsewhere in this chapter. The clock maintains information about the date and time of day, this is used in some applications to date stamp messages. Communication between the token and the outside world is by way of a three-wire serial interface.

Communication between the two processors takes place over an 8-bit bidirectional bus buffer constructed in the semicustom chip. A data block containing the message, exponent and modulus required for an RSA calculation is sent to the RSA processor via this interface, the result is returned in a similar way. Block transfers are completed in a few microseconds, a time not considered significant when set against the calculation time.

The token contains secret information unique only to itself. No other token contains the same secret so the compromise of one does not put the security of the whole system at risk. Nevertheless, measures must be taken to detect tampering and destroy the secret information upon detection. Possession of the secret key would enable an intruder to falsify

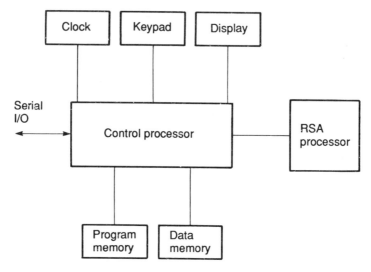

**Fig. 6.3** NPL token block diagram.

transactions in which the token was engaged. The tamper resistance built into the prototype token simply detects when the case is opened and destroys the secret data by discharging the RAM chips. Although adequate for our prototypes, this form of tamper detection is nothing more than a gesture and new circuitry has been designed to perform these functions properly. We have indicated elsewhere the important functions performed by the display and keyboard; clearly these devices must also be included in the tamper protected area. It is necessary to note that tamper resistance will add significantly to the cost and size of the token.

The speed required of the RSA calculation implies a device that consumes a lot of power. It is unlikely that this can be supplied conveniently to the token without direct electrical contact so the token takes its power from the terminal into which it is plugged. RSA calculations are only required when the token is communicating with a terminal, so it could be supplied separately; other functions such as a calculator, clock and a database would then be run by the TMS7000 under battery operation.

The implementation of the RSA algorithm on the signal processor is capable of carrying out one full operation of the algorithm in about 3 seconds. This implementation has been described by Clayden [5]. An increase of speed has been achieved by using the method [6] based on knowledge of the two primes that make up the public modulus. Knowledge of these primes is permissible in the device that holds the secret key. In other words the two primes method can be used to generate signatures, but not for checking them. By this means it has been possible to reduce the algorithm time for signature generation to about 1.4 seconds. For signature checking it is possible to gain a substantial increase in speed by limiting the public exponent to, say, 17 bits. This places no restriction on the size of the secret exponent and does not reduce the security of the signature operation. The speed of signature checking is not exactly inversely proportional to the size of exponent because of overheads in the computation, but a very substantial reduction in checking time is possible. In our implementation, signature checking using this method takes 60 ms.

## 6.5 MINIATURISATION

It is possible to argue that a token carrying display and keyboard should not be reduced in size to that of the standard plastic card as has happened with the smart cards. Very many people now carry small pocket calculators as a matter of course and the intelligent token could easily be reduced to a compatible size. Smart cards have not yet been in wide enough use to allow a judgement to be formed as to their durability. If the ISO draft standards for physical properties of smart cards are to be regarded as firm guidance, then cards are going to need the ability to withstand very severe handling. A small calculator-like object can be made much stronger. We understand, of course, the desire to make the smart card compatible in dimensions with the existing magnetic stripe card standards because of the volume of installed equipment accepting the latter.

Of the 11 chips in the current token, 6 are memory devices and would be the first target for chip count reduction. Integrating these devices into the processor would also increase the security of the token. A low power processor could be permanently powered by battery; under these circumstances it could also act as the time of day clock. A two chip token could be built today using semicustom microcomputers now available. A single chip design using the latest digital signal processors is a distinct possibility in the near future.

## 6.6 BIOMETRICS

It should be plain from the introduction to this chapter that we are not satisfied with the level of security achievable in systems designed around personal identification numbers. We have mentioned the possibility of replacing PINs by biometric methods. Because the NPL token is the size of a small calculator, it is a realistic aim to mount a suitable interface for biometric identity verification on the case of the token; this could be based on written signature, fingerprint, vein scanning, etc. It is convenient that the type of processor required for realising the biometric measurement is just the same type that has been used for the RSA implementation in the NPL token, a signal processor, such as the TMS32010. If a biometric identity verification method is to be implemented in the intelligent token, further work is required and this depends on achievement of acceptable performance in the chosen biometric identification method. It is unrealistic to expect that any biometric method will be totally error free and the overall transaction system design will need to take this into account. When a PIN is presented to an identity verification system there is no margin for error − the PIN is either correct or it is not; in a biometric system a degree of tolerance must be built in to allow for variability in measurement. Because of this degree of tolerance, a higher level of security may be achieved by retaining the PIN as a supplementary check on identity.

One method currently of interest relies upon the unique features in the blood vessel patterns on the back of the hand. This development, called Veincheck, is under investigation on behalf of the British Technology Group. If built into the token it might consist of a row of four infrared emitter/detectors built into the base. Identification would be performed by wiping the base of the token over the back of the owner's hand before confirming a transaction.

### 6.7  FUTURE DEVELOPMENTS

Clearly, given demand for the device, it will be feasible to miniaturise the intelligent token still further, possibly even down to the compass of an ISO standard plastic card, yet carrying keyboard and display. Products of this type are already being designed by various manufacturers, but not making use of the signature capability of the NPL device. A smaller device would undoubtedly be more convenient and, therefore, possibly be more acceptable to users. Adoption of ISO card dimensions would ensure compatibility with magnetic stripe card readers adapted to take intelligent tokens. However, a very small size is not necessarily an advantage. The durability of a very small design of this nature is not yet tested. User habits with plastic cards are notoriously bad and a very small token would have to be resistant to considerable mishandling.

Even more critical than the size of the device is its cost. For mass penetration of a market low cost is essential, especially if the competition is provided by cheap magnetic stripe plastic cards. If the cost of intelligent tokens remains high, then their use may be limited to applications where cost is not so important.

If it becomes possible to achieve a sufficiently low unit cost, then mass markets may become available and the token used for transaction control and payment for a wide range of services in many environments, such as in the home, in public utilities (transport, telephone, etc.), the office and in shopping. If the amount of internal storage can be increased significantly, the token applications can extend to serving as a personal memo, communicating with a database held on a personal computer. It is also possible to envisage applications in the medical field, with personal medical records held within the token, and prescriptions issued by the doctor's terminal directly to the token and the pharmacist's terminal reading them securely. Because many people are quite happy to carry small calculators on their person, the addition of calculator functions to the token might be an attractive option. Addition of such a capability might overcome the problem of token cost. People willing to buy a pocket calculator might be only too willing to pay a little extra for the substantial addition in capability.

In this chapter we have attempted to point out the need for an identity token with greater security than the magnetic stripe card. We are also concerned that terminal security in smart card systems may be a weakness. The intelligent token has the advantage of greater safeguard for PINs, whilst offering the further capability of ensuring the integrity of transaction messages. The cost of providing terminal security can be reduced by introduction of the intelligent token, because the terminal need not contain protected secret parameters.

## REFERENCES

[1] Chorley, B. J. and Price, W. L. (1986) 'An intelligent token for secure transactions' *Proc. IFIP/Sec'86*, Monte Carlo, December 1986, 442–450.

[2] Price, W. L. and Chorley, B. J. 'The intelligent token or 'super-smart' card' *Proc. SmartCard 2000*, Vienna, October 1987 (proceedings to appear).

[3] Commission of the European Communities (1986) Open Shops for Information Services (OSIS), Final Report, Cost Project 11 ter, CEC, June 1986.

[4] Rivest, R. L., Shamir, A. and Adleman, L. (1978) 'A method of obtaining digital signatures and public key cryptosystems'. *Comm. ACM.*, **21** (2) 120–126.

[5] Clayden, D. O. (1985) 'Some methods of calculating the RSA exponential' *Proc. Int. Conf. on System Security*, Online, London, October 1985, 173–183.

[6] Quisquater, J.-J. and Couvreur, C. (1982) 'Fast decipherment algorithm for RSA public-key cryptosystem' *Electronic Letters*, **18** (21) 905–907.

Chapter 7

# Automated Personal Identification Methods for Use with Smart Cards

## JOHN R. PARKS

(J R P Consultants)

*The PIN is insecure. Biometric measurements on card holders are needed to confirm their rights of possession.*

## 7.1 INTRODUCTION

The validation of a claimed identity or claim to represent some third party has historically depended on one or more of four characteristics of the person presenting themselves. These are:

(1) some recognisable thing that he has, e.g. a key, a ring, a personal sign (signature), a card (magnetic, smart or otherwise), etc. In short a token.
(2) some piece of knowledge known only to authorised persons, e.g. a password (PIN), intimate personal detail (mother-in-law's sister's car registration number, or something simpler!), etc.
(3) some inimitable characteristic of a physical nature, e.g. fingerprints, facial features, hand geometry etc.
(4) some characteristic behaviour, e.g. speech pattern, gait, writing dynamics, etc.

### 7.1.1 Tokens and things possessed

The problem with using tokens to validate a claimed identity is that such tokens can be obtained by impostors, by finding, stealing, coercion, deceit or collusion. According to the uniqueness and technology involved tokens can, more or less easily, be copied or counterfeited. A token once 'detached' from its rightful owner has clearly compromised the security system which depended on it. Unless widely duplicated the risk is however limited to the new holder.

The credit card is the best known and most widely used token today; recognising its vulnerability the card is often supported by 'something known' i.e. the ubiquitous Personal Identification Number (PIN). The snags with PINs are well known though they are still the preferred method for cash access purposes at autotellers. In a complex modern life any individual has several cards each with a different associated PIN. The problem of remembering the number which supports each card is significant and often (frequently) solved by writing down the number or a simple coding of it and storing it in close proximity to the card. There are documented accounts in the USA of a large proportion of recovered 'lost' cards having the associated PIN written on them.

The advent of the 'watermark' magnetic card made 'skimming' (transfer of data from one to one or many other cards) more difficult as the card stock is validatable and very difficult to counterfeit, though this would not immediately prevent collusive use of two or more watermarked cards. The so-called 'smart card' or 'IC card', the central topic of this book, brings a further dimension to card tokens, namely their inbuilt and tamper resistant 'intelligence', and also the extreme difficulty of counterfeiting them.

However, no matter how intelligent a token is, it is still independent of its valid user and can be 'obtained', by fair means or foul, and is dependent on associated knowledge in the user to 'activate' it. Such knowledge can similarly be 'obtained'.

### 7.1.2  PINs and other knowledge

In a face to face interrogation the bona fides of a person presenting himself and claiming a particular identity can be tested by cross-questioning him on the basis of personal details including knowledge of specific facts which the inquisitor knows that the presenting person also knows. If taken to sufficient depth and in time then such methods can be secure and impostors detected. However their use in a day-to-day commercial context would be unrealistic where speed is generally of the essence. Some recent work in Japan has attempted to mechanise this Q/A interrogative approach [1].

Secure commercial techniques have been developed for validating the authenticity of (coded) messages; systems of numeric 'passwords' have been devised and encryption techniques for protecting communication channels are well known, these are described elsewhere in this book. There are also simpler methods based on the knowledge of a simple four or five digit number to support a presented token, the ubiquitous Personal Identification Number (PIN) or its counterpart for access control the Personal Access Number (PAN). These are only useful to protect modest value because their intrinsic security is not very high.

As already mentioned, the practical problems of remembering several multidigit numbers are often overcome by writing the numbers down and storing this record close to the relevant cards. Banks are still very dependent on simple PIN systems but have eased the user's memory problem somewhat by allowing him to invent his own number. This eases one problem but creates others in that numbers invented are likely to relate closely to the individual (e.g. birthdate (self or spouse), telephone number, etc.) and to be used as 'universal PIN numbers' for all the cards held by that individual. Non-random numbers are likely to be more easily guessed or observed; universal numbers once 'cracked' then compromise all that individual's cards.

There is therefore a place for automatic techniques to verify the presenter of a token as well as to verify the token. The rest of this paper concentrates on this possibility with no further comment on the design of tokens or of the use of 'knowledge' to support them.

### 7.1.3  Automatic Personal Identification (API)

The topic of automatic verification of claimed identity or of direct recognition of persons has been current for at least 15 years, for example, palmprints (Palmguard 1969), fingerprints (IBM 1965), speech (IBM 1967), and signatures (Connetta 1969)).

After several years research commercial products have emerged during the past decade, e.g. Identimat (1978), Qsign (1982). More recently a spate of products have appeared and an API vendors' trade association has emerged in the USA.

Several *signature-based systems* using dynamic signatures, i.e. using the dynamic as well as pictorial nature of a signature as it is written, have been developed (Quest, AITSL, Signify, Confirma). There are also some systems for verifying 'cold' or static signatures for use in for example, clearing cheques for payment (ROCC, AutoSig). IBM have been working on a dynamic signature verification system continuously for some 15 years and are widely rumoured to be releasing a product 'soon'.

*Fingerprint systems* for commercial (Fingermatrix, Thumbscan, Identix) or forensic (De La Rue Printrak, NEC, Morpho) purposes are available. *Retinal patterns of blood vessels* (EyeDentify) are used for high security low throughput access control. *Hand geometry* (Stellar (IDENTIMAT), Recognition Systems Inc.) and palmprints (Palmguard) find favour for mass access control. *Keystroking rhythms* (BioAccess) at the keyboard

offer a method for continuously verifying the user of a terminal. Much
has been written about *voice-based systems* and some useful demon-
strations have been staged (though generally for security applications and
behind closed doors); two commercial products are available (ECCO,
Votran). *Facial recognition* continues to be researched but, so far, no
commercial system is yet on offer.

### 7.1.4 Anatomy, and performance specification and measurement

Before describing individual approaches and methods of API it is useful
to outline the anatomy of any API system in schematic form (Figure 7.1).
   The general form begins with an input transducer to capture the property
being used for verification; i.e. a microphone for speech, xy digitiser for
written signatures, video-camera for fingerprints etc. The transducer output
signal is then subject to signal analysis. After analysis the signal changes
for the process of enrolling new users or for live-use with previously
enrolled users. The enrolment algorithm manages the collection and
storage of reference inputs from the enrollee. In live-use these references
are used in a comparison process to assess the mismatch between an
offered specimen and the stored references. A decision mechanism then
either accepts the specimen against the references, requests another specimen, or, after repeated unsuccessful attempts rejects the user. This decision
mechanism is often part of the host-system, this is discussed further
below.
   Clearly before a potential user of an API device can proceed he has to

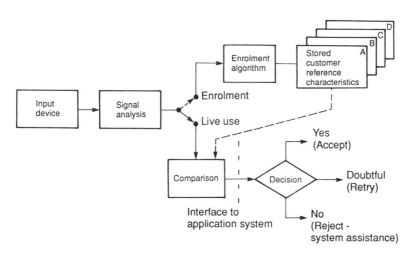

**Fig. 7.1** Biometric system schematic.

be admitted to the system through an enrolment process, subsequent live-use of the system then refers to the results of this enrolment process. Effective management of the security of the enrolment process is mandatory if the total system security is to be of a high order.

Thus all API systems have two aspects to their operation, namely enrolment, and live-use. The enrolment process requires the collection of one or more certified examples of the verification phenomenon being employed (speech sample, fingerprint, signature, or whatever) to be provided by the person being enrolled. These reference inputs can be stored individually (fingerprints) or, more generally as the statistical properties of the provided set of nominally identical inputs. These references are stored with other user information − possibly within his 'smart card'. In live-use these references are recovered upon the user claiming an identity, and he is then challenged to reproduce the action used to generate the references. Comparison of this specimen with the stored references yields a value indicating the degree of mismatch of the specimen.

If the mismatch is small enough then the user can be allowed the facility being sought, or he can be refused if the mismatch is excessive. Refusal would generally be the result after allowing one or more further unsuccessful attempts to produce a satisfactory match with additional specimens.

Assuming that the application being protected through the use of an API (sub)system is one for monetary value transfer then the degree of mismatch judged to be tolerable is modified by such system considerations as the value of the transaction, the value of the client and other aspects including the history of attack of that user's 'account'. It follows that this judgement has to be made by the host system based upon the quantified assessment, by the API device, of the mismatch existing. For simpler tasks such as access control where only an unmoderated decision is possible the decision threshold would normally be incorporated in the API device.

In discussing performance of the various API devices there are two characteristics which have to be specified. These are the type 1 or *false rejection* error rate (i.e. the 'insult rate' when the genuine user is refused); and the type 2 or *false acceptance* error rate (i.e. the 'penetration' rate when an impostor is accepted as genuine). These characteristics are illustrated in Figure 7.2. It is clear that the two error characteristics are opposed, and that one can be reduced at the cost of increasing the other. This trade-off is effected by adjusting the tolerance setting applied to the mismatch value (and also by the number of retries allowed if first attempts fail). The curves shown are for illustrative purposes only and do not represent the detailed behaviour of any particular device. Apart from the two classical error rates indicated there is a third characteristic which has significance in API devices based on user behaviour, such as speech or

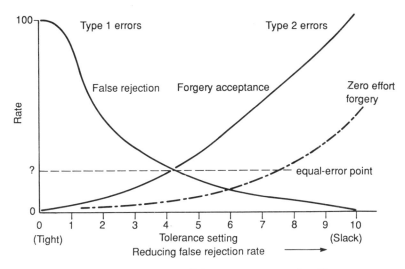

**Fig. 7.2** False reject/false acceptance trade-off.

written signatures; this is the *zero effort* error rate obtained when an arbitrary action is offered rather than an explicit attempt to forge or mimic a target object, i.e. an arbitrary spoken phrase or scribbled writing.

Measurement of the performance of API systems is a subject fraught with problems, in common with most other man-machine interaction situations. In this particular context the usual method of evaluating perform- ance is through user or field trials. While a well conducted field trial, hopefully managed by an organisation independent of the supplier, can give a good indication of the performance obtainable in the specific ap- plication tested, the validity of the results in any other situation with another user population is questionable. The problems of comparing dissimilar devices in different applications with a different user population are not well understood. There have been attempts to take a statistical approach in estimating the achievable performance of a given system. This is an obvious ploy in measuring the change in behaviour of a developing experimental system using the same database of input 'signals', but in spite of its more scientific basis such techniques are still dependent on a user situation to validate the model and cannot accurately reflect the change in performance due to change of application site, user population motivation of users or the control of the experiment. Caution must therefore be exercised in comparing quoted performance figures. Com- mercial pressures can lead to overly optimistic extrapolation from very limited (supplier) trials.

In the USA there have been several comparative exercises (MITRE Corp., Sandia Labs) aimed at determining the performance of (near) commercial systems for possible use by US armed forces. These are the most objective figures available but are limited to what was judged to be the state-of-the-art at the time the tests were mounted. This topic is explored further later in this paper.

There is then, much activity in the laboratories and a range of products available.

### 7.1.5  Product development

It is no surprise that the major efforts in this field, judged by the volume of literature in evidence, are to be found in the USA with major firms such as IBM, Rockwell, TRW, Sylvania and Westinghouse much in evidence plus a plethora of small entrepreneurial companies, some spun off from institutions such as SRI and TRW. Activity is also evident in the UK, Japan, Canada, Israel, France, West Germany and Scandinavia. European firms have acquired some US technology – De La Rue Printrak exploits Rockwell International technology and Siemens have bought out Threshold Technology Inc. Over a hundred firms, institutions and government agencies worldwide have been identified as having or having had substantial activity in the topic of API (see section 7.8).

For commercial application of API devices in other than specialist applications it is essential that the devices be user friendly, offering little offence or insult to the user through rejecting him, and that they be economic, in most instances this means very cheap if the devices are to be used in large numbers in support say of ATMs.

While the processing power to support a particular technique is relatively cheap and abundantly available, the input transducer is often complex and hence expensive. For instance the input transducer for a facial recognition or fingerprint system is in essence a TV camera and associated optical components which are relatively expensive; the input microphone for a speech-based system is, on the other hand, cheap.

There are also social niceties which have to be observed in a system for use by the general public (or valued customers). The authorisation of a cash debit by signature is conventional and unexceptionable; the use of a fingerprint with its overtones of criminality may not be received well.

### 7.2  PHYSICAL FEATURES

This group of API methods exploit the fact that all people have a variety of uniquely identifying features in their physique. Personal recognition of known individuals through their characteristic stance or facial appearance is a common experience. Their use is, and has been for many years, a basis for proving identity. Fingerprints are a classic example, though in passing, it is worth commenting that the forensic use of fingerprints is used at least as much to break aliases as to identify miscreants through scene-of-crime prints (partials). Other forensic methods regularly use body fluids, hair and other material samples, and dental state. 'Mug shots' are used in police records and on identity cards of all sorts whether for claiming advantageous travel fares or on formal documents such as passports and other official instruments.

Forensic methods are not generally adaptable to commercial use on a day-to-day transaction basis and will not be pursued in any depth here. All such systems depend upon human interpretation which often includes a variety of other inferences about the suspected intent or purpose of the individual being investigated. Automatic systems have no such secondary information unless two or more methods are explicitly combined.

The variety of physical characteristics which have been suggested and investigated for automation is extensive. The list of candidate characteristics include.

(1) Facial features, full face and profile
(2) Fingerprints
(3) Palmprints
(4) Footprints
(5) Hand geometry (shape)
(6) Ear (pinna) shape
(7) Retinal blood vessels
(8) Striation of the iris
(9) Surface blood vessels (e.g. in the wrist)
(10) Electrocardiac waveforms.

### 7.2.1 Facial features

There are a variety of systems for constructing facial images in use by police authorities e.g. Identikit, SIGMA/IRIS, [2] but there are no known automated systems in or near commercial products which base personal verification on the analysis of facial images.

There are however, significant activities evident in the literature of attempts, largely in academic institutions, to devise methods for describing faces. The activity is purely scientific and aimed at developing the basic (AI) techniques for deriving usable facial descriptions. The best results reported under rather favourable conditions, using 121 subjects, are from Case Western Reserve University [3] who claim a false rejection rate (FRR) or Type 1 error rate of 4% and essentially zero false acceptance rate (FAR) or Type 2 errors.

### 7.2.2 Finger and palmprints

The literature on finger and palmprint technology is, curiously, largely confined to patent documents which contain little if anything on performance attainable. Most of the material relates to analysis of fingerprints for forensic purposes and is usually only semiautomatic, i.e. a man aid. Firms which clearly major in forensic systems include: Fingermatrix (active from 1970 to date), Sperry Rand (active from 1973 to 1978), Rockwell International (work, products and further development now taken over by De La Rue Printrak, active from 1976 to date). There were also a number of short lived activities in the late 1960s to early 1970s in the USA and Europe in the early days of development of pattern recognition technology often using optical processing – many salutary lessons were learnt at this time! Late arrivals on the scene include Logica and CAP (1985, both supported by the UK Home Office), Siemens 1980 [4], Fujitsu 1985, NEC 1980 [5] and a major exercise by NBS between 1982 and 1985 [6].

Substantial academic work has recently been published by Isenor and Zaky, 1986 [7] on topological-graph representation of fingerprints in which all individual ridges are characterised with their topological relationships to adjacent ridges. This work appears to be the most thorough going approach of recent times with a real chance of overcoming problems due to distortion of the prints associated with geometrically matching spatially distributed minutiae. The graph representation of a significant fraction of a fingerprint is likely to be very extensive. Regrettably the authors give no indication of the computing power needed to implement their approach, either in volume or run-time terms, though they are clearly substantial. It is therefore a contribution to the forensic interests in fingerprints and not easily adapted for low cost requirements.

De La Rue/Rockwell are the largest supplier of forensic systems – some 35 out of 40+ installed; NEC is in second place. These highly expensive forensic systems will not be discussed further here.

Entrants to the (much) lower cost API device development and market include:

- Sibany Mfg Corp, 1966 [8]. An integral geometry approach of supreme optimism.
- Stellar Systems 1983 [9]. A primitive system for comparing local patch (density) of a certified and test fingerprint.
- Identix 1985 [10]. Have developed an optical system for scanning a curved (270°) longitudinal area of a fingertip.
- Fingermatrix 1984 [11]. Direct comparison of small areas of prints by correlation under scanned translation to accommodate misplacement and distortion.
- ThumbScan.

Of these, Fingermatrix, Thumbscan and Identix are currently offering products.

Palmprints have been explored as an alternative to fingerprints. An early system proposal from IBM in 1977 [12] used mechanical 'probes' to determine the 2D surface contour of a hand. Palmguard have been active since 1971 [13] and have products on the market. The palm offers not only a much larger test object but also a system of crease lines which are at once simpler to analyse and less in number than the ridges of a fingerprint but are acceptable as sufficiently characteristic of individuals for the purpose of API.

### 7.2.3   Hand shape

The product IDENTIMAT currently available from Stellar Systems Inc., San Jose, Ca. [14] has been on the market from changing sources since 1978. It has probably the biggest installed equipment base of any method of API, with over 300 units in the field for up to 7 years. Such literature as there is implies an equal error rate (i.e. FRR = FAR) of the order 0.1%.

The Identimat product was generally used for access control though it is known that the US Army trialed the device in support of pay cheque encashment security (against impersonation) in 1984.

Mitsubishi Denki appear to have entered the field in 1983 [15] with an overhead illuminated hand silhouette system. No significant details are available on the Mitsubishi device.

### 7.2.4   Ear shape

The shape of the pinna (external ear structure) is seen to vary between individuals and could perhaps form the basis for a PI technique, but little if anything is known about the uniqueness of pinnae shape. Methods which suggest themselves for testing the shape of individual pinnae are optical image processing and acoustic measurements. There are from time to time conjectures that the internal ear generates sound as part of the action of sound analysis (parallels with optical holography?). The most recent conjecture appeared in *The New Scientist* 1983 [16]. No work on this possibility is evident.

### 7.2.5   Retinal/iris patterns

The analysis of patterns of blood vessels on the retina has been evident since 1977 when intellectual property was registered by Eye-D Development II Ltd, and finally manufactured by EyeDentify Inc in 1985 [17].

The EyeDentification System 7.5 access control system verifies users by scanning the pattern of blood vessels of their retina. This pattern is claimed to be stable after infancy and unique to each individual (in the same way as a fingerprint?). An optical system using a low powered infrared LED scans an annular ring of the retina centred on the fovea (alignment is achieved by requiring the user to focus on and align two sighting marks). The analogue signal from this ring is reduced to a binary form and a code of 40 bytes (320 bits) generated (indicating an angular resolution of the order 1 degree). The earlier patents on the device describe two concentric scans centred on the fundus of the eye where blood vessels enter the eye but the active participation of the user eases the problem of locating the fundus automatically.

False acceptance rate is claimed to be 0.0001% and false rejection better than 0.1%. It is not clear how the false acceptance error rate was measured. Clearly to prove a performance of parts in a million several million verifications have to be performed; it is understood that the figure is obtained by theoretical extrapolation from field trials.

This device requires considerable skill/training of the user and is clearly intended for high security closed population access control where it is highly successful, though expensive.

In the search for new methods of personal identification it is conjectured that the striations of the iris are essentially unchanging with time for any individual and vary widely in their pigmentation, intensity and structure between individuals. No work is in evidence on this possibility.


### 7.2.6  Electrocardiac waveforms

Work on this technique is hinted at by SRI as part of a multiclient project proposal. No information is forthcoming, even as to whether the project is being pursued or not. It is known that individuals do show variation in their ECG traces but this variation between individuals is likely to be confused by the level of activity of the individual immediately prior to being tested for PI purposes. While instrumentation exists for detecting and first order analysis and monitoring of ECG, this approach is highly conjectural for API use on a routine basis.


### 7.2.7  Surface blood vessels

A device recently patented by BTG [18] describes methods for obtaining a one dimensional scan of blood vessels in the wrist or across the back of the hand. It is contended that such distributions are characteristic of the individual. As in the retinal blood vessel system there appears to be no universally prescribed position for blood vessels though it is assumed that whatever pattern is established is fixed for a substantial period of years. Subject to changes due to thrombosis or other mechanical disturbances this seems intuitively reasonable but has not, so far as this author is aware, been subjected to scientific tests.

## 7.3  BEHAVIOURAL CHARACTERISTICS

This group includes individual behavioural characteristics. These are characteristic actions which are highly learned, are performed more or less unconsciously and are much practised. They are real-time in nature and therefore include a measurement period during which the behaviour of interest is monitored. Being behavioural in nature the methods used have to contend with variation in the individual's behaviour which may be affected by environmental conditions and distraction as well as the state of health of the individual. Behavioural functions include:

(1) Speech
(2) Writing
(3) Gait and Grasp
(4) Keyboarding

### 7.3.1  Speech parameters

The use of speech to establish user identity has excited more interest than any other form of user verification (also some activity for identification of speaker in order to aid speaker independent speech recognition).
    Applications are split into two major classes, namely:

(1) forensic systems to establish identity of suspects in criminal cases, e.g. kidnap case phone calls, blackmail etc.
(2) automatic personal identification systems for use in e.g. banking, access control etc.

Forensic methods, 'Voice-print' and the like, are at best semiautomatic and generally implemented in laboratory situations. Such techniques are controversial and have been, and continue to be, subject to substantial attack by phoneticians and other scientists for their lack of scientific base [19, 20, 21]. Though interesting, this subject is not relevant here and will not be pursued. There is also some work evident on detection of stress (e.g. lie detection) through voice tremor [22].
    Speaker identification and verification has received substantial attention by Texas Instruments [23], Bell Laboratories [24], Threshold Technology

(now a subsidiary of Siemens) [25] and Philips (Hamburg) [26, 27, 28]. Significant activity and principal results reported date from mid to late 1970s. Some product development has been evident in the 1980s (Identicator, Votran (using TI technology), Computer Gesellschaft Konstanz (i.e. Siemens).

Other firms and University Departments are also in evidence but without much depth. These include IBM [29], Perkin Elmer [30], NCR [31], Purdue University [32], Tokyo Shibaura Denki [33]. Publications by Westinghouse [34] and University of Windsor, Ontario [35] are of particular interest as they explicitly address the problems encountered in the use of noisy and distorting channels − dialled lines. Recently workers at Carnegie-Mellon University [36] have shown techniques for context free speaker recognition. This is significant as it potentially offers a method for continuously verifying the user of a speech circuit, this may be necessary to detect impersonation.

All techniques (except C-MU) rely on the use of a nominated utterance as the test object.

The four principal firms in this field have adopted different approaches;

• TI based their work on characterisation of (vowel) segments in monosyllabic four word phrases.
• Bell used temporal features derived from a sentence-long utterance (12.5 s).
• Threshold Technology/CGK used a technique based on formant features of selected words.
• Philips used long-term spectra over complete utterance (12 s)

The work at TI and Bell appears to be contemporary with the 'Speech Understanding Project' mounted by DARPA in the early 1970s and no doubt gained much impetus from that (over) ambitious project.

Most of the work reported has employed filter banks to perform initial analysis of the speech waveform, other possible techniques staying entirely in the time domain such as (multiple) auto-correllation have not been reported for PI; they comment on the convenience of filters plus rectifier and lpf for the front-end. Bell used linear prediction techniques in their early work.

Several of the experiments use non-linear time warping to align successive utterances from users; this rather expensive technique may give way to more easily computed hidden Markov models now used in the speech recognition world [e.g. RSRE, 37] but this has not yet penetrated to speaker recognition/verification.

While more work is evident in speech than for example signature methods, the latter report much more thorough field trialling; quoted results from trials are comparable for the two though speech systems appear to be more vulnerable to mimics (let alone tape recorders) than signature-based systems.

In comparing the performances of different speech systems it is again necessary to note that tests are performed on different populations under different circumstances; it should also be noted in comparing speech systems with others, particularly signature-based systems, that the simple impostor attack is not the same as a forger's attack but corresponds to the so-called 'zero-effort forgery' in which other user's (normal) signatures are used to attack a specific user. In a speech-based system the attack by mimics is the appropriate attack corresponding to a forger in signature systems. A more detailed description of published work on speaker verification is given in the appendices to this paper.

### 7.3.2 Writing dynamics and statics

We must distinguish between the two possible signature presentation situations, namely static and dynamic. The appearance of a signature on a document in the absence of the signer (as in an authorised cash transfer) is said to be 'static' or 'cold' and the signature can only be treated as a two dimensional image. In the second situation the signer is present and is required to write a normal signature on a prepared surface, or with a designated stylus. This surface or stylus measures the motion and pressure of the writing stylus in real-time. This 'dynamic' situation is under real-time direction by the system.

*7.3.2.1 Static signature verification*
There have been a number of systems and products for automatically comparing stored facsimiles of individual signatures with a presented specimen, for example, a bank teller's desk or for a centralised cheque clearing operation [38, 39].

The security actually achieved with static signatures must be open to question as it is a trivial task, given a modern photocopying machine, to copy a signature from one document to another.

*7.3.2.2 Dynamic signature verification*
The user writes with a particular apparatus for the act of signing. This may be either an instrumented pen (as preferred by IBM [40] and SRI [41] for example) or an instrumented platten on which the document to be signed is laid during signing, e.g. [42] or both a special pen and a platten which interact (e.g. Signify, AITSL).

Instrumented pens variously measure the force applied to the stylus (e.g. SRI) and/or the acceleration of the pen in two or three orthogonal directions including the pen axis (e.g. IBM). The instrumented platten is typically a digitiser which is either pressure activated (Quest) or couples electrically or magnetically with the pen (Signify, AITSL). Most platten devices measure the absolute position of the pen at frequent intervals and yield an accurate and continuous representation of the position of the pen during writing.

Instrumented pen systems are responsive only to the first or second derivative of the pen's motion (velocity or acceleration) and are therefore not capable of absolute positional measurement. This may not be significant but only the digitiser approach gains enough information about the signature to reproduce it graphically and allow absolute spatial measurement of the writing . The instrumented pen approach can measure variation in writing pressure (rather than simply pen-up or pen-down) and this may compensate to some degree for the loss of positional absolutes.

It is, of course, possible to combine the digitiser approach with mechanisms in the pen or the platten to also measure writing pressure. Such has not been used, presumably for cost reasons.

The instrumented pen approach has the advantage that users will always use the particular writing instrument supplied by the system, but even with an estimated manufacture cost of $50 (IBM) the cost and interruption of service if an instrumented pen is 'lost' or damaged is an obvious hazard and effectively excludes the use of such an approach in unsupervised situations.

The digitising platten on the other hand, has to cope with whatever instrument the user chooses; though the user is generally motivated to achieve whatever service is being offered and will probably be readily persuaded to use a preferred 'hard' writing instrument such as a ballpoint pen.

A number of low cost devices have been proposed for extracting information on the dynamics of writing. For example:

- The velocity of a writing stylus can be assessed as it moves over a document resting on a ruled (ridged) surface which includes a microphone in either the pen [43] or the surface [44] to pick up the resulting 'noise' of writing. A pressure sensitive device measuring velocity in a single dimension (i.e. vertical) intended for inclusion in a POS terminal has been proposed [45].
- Use of styli incorporating accelerometers is reported fully by IBM [46].
- Instrumented pens measuring the three orthogonal forces (vertical, horizontal and pressure) induced in writing have been reported extensively by SRI [41] and others [47].
- Techniques for analysing writing pressure alone are reported; binary pen-up and pen-down signal [47] and continuous pressure [48].
- Several techniques are described which exploit magnetic coupling between pen and platten [49].
- Acoustic radar has been used at ETH, Zurich to track the motion of a stylus [50].

Descriptions now follow of several methods developed for dynamic signature recognition which have been pursued and reported in substantial detail.

There are two fundamentally different approaches to writing-motion analysis, namely those which attempt to match the analogue waveform(s) produced with known reference waveforms (possibly exploiting the natural segmentation of the waveforms due to pen-down or pen-up state changes), and those which attempt to determine overall, holistic, measures of a signature for statistical matching with reference values.

These two approaches will be treated separately.

*Waveform matching*      Both ETH and particularly IBM have explored this approach over the last decade. The following remarks outline a well documented experimental situation devised by IBM and described more fully in the appendices. Further work by Lew also of IBM [51] indicates further benefit to be gained from a more thorough understanding of the kinematics of writing (but requiring more instrumentation of the pen to fully determine all six dimensions of motion).

The IBM approach uses an instrumented-pen [52, 53] which contains two accelerometers and a tip-pressure transducer. The accelerometers are mutually orthogonal and measure acceleration normally to the axis of the pen. The pen is of cylindrical form thus making no demands on the user with respect to orientation. As the attitude of the accelerometers is thus undefined in use the signals from them are combined during a (software) signal conditioning process, to indicate acceleration with rotational invariance but retaining sign. The pressure transducer, like the accelerometers, is piezo-electric giving an analogue output, and because of its rigid structure imparts no additional 'feel' to the user (as is the case with mechanical switch mechanisms to detect pen-down).

After initial signal conditioning the analogue signals are sampled at 80 Hz into eight bits per sample per signal channel.

In the most recent (and detailed) description of the IBM approach [46] the algorithms used for comparing signals from a given signature with those from a nominated target for verification are presented. All processing is on the basis of waveform comparison, taking each segment individually, with adjustments for segments of very long duration.

Signal analysis begins with examination of the pressure signal; segments are separated and used as a basis for all further processing and comparison. Segments extending for more than 0.7 s are artificially segmented and components marked as coming from the beginning, end, or middle of the large segment.

Assessment of the similarity of an offered signature to a reference is based on a set of five measures:

M1 Segment alignment

M2 Pressure correlation

M3 Acceleration correlation

M4 Pressure coherence (based on the spectral density of the pressure signal)

M5 Acceleration coherence (using spectral density of acceleration).

These are defined fully in the appendix.

For M1 it is necessary that the number of segments are the same in specimen and reference; this condition is not likely to be met reliably from one signing to another. Accordingly individual segments in the specimen may be merged (adjacently) or divided in such a way that M1 is minimised. (This is done before any long segments are split up).

The Similarity Measure, SM, between a specimen and a reference signature pair is computed as:

$$SM = \sum_i Sm_i$$

Where $Sm_i$ is calculated as following:

$$Sm_i = \frac{(M_i - MID_i)^3}{(HI_i - MID_i)^3} \qquad M_i >= MID_i$$

$$Sm_i = \frac{(M_i - MID_i)^3}{(MID_i - LO_i)^3} \qquad M_i < MID_i$$

in which $MID_i$, $LO_i$ and $HI_i$ are chosen by a mechanism which reflects the expected values of that measure when applied to a body of genuine and forged signatures. See reference for more details.

These formulae yield an index which indicates the degree of match between the specimen and the reference signatures. A threshold value applied to this index then adjudicates acceptability or otherwise of the specimen.

References for each individual user are established by a set of two more signatures which are collected from a user at enrolment.

From these signatures two are selected as first and second references by computing the SM for each signature against every other signature in the set collected. Find the minimum of these maximum values. Finally select the two signatures that have the maximum of the hand found minima.

This technique is claimed to detect and accommodate users who present two signature styles, but can be confused by more than a single maverick (e.g. incomplete or otherwise damaged) signature. If a maverick is detected it is discarded and a new pair selected from the reduced set.

In a practical enrolment situation some six sample specimen signatures are collected. Apart from the first and second references a 'bucket' of four more signatures is maintained which contains the four most recently accepted signatures so that a set of six is always available from which the two references can be selected. This mechanism ensures that user's gradual changes in behaviour with the system are accommodated.

The decision mechanism is based on the SM values obtained from the first and then second reference or if either is only marginally rejected then the signatures in the bucket are also used.

In order to increase the protection against successful forgery and to reflect the difference in the degree of variability between users a number of individual parameters are determined at enrolment (and subsequently updated) for each user and are stored with his references. Namely, the length of quiet sections for lift-off detection; the maximum segment length; the amount of translation needed to obtain acceptable correlation, and some ad hoc factors determined empirically.

If an initial specimen signature is found unacceptable then one or two more samples are requested. If they also fail then the user is rejected.

In discussing the measurement of performance IBM note that all users should receive a similar level of service however variable their signature and define two further useful factors to characterise overall system performance. These are a utilisation index indicating the cumulative percentage of a user's particular performance, and percentage of users with zero errors; again detailed in the appendix.

During Autumn 1983 a trial was undertaken with a group of 108 'manufacturing employees' in IBM's plant in Charlotte, NC. After enrolment, users were asked to sign twice daily for nine weeks. A signing session consisted of up to two attempts while standing at a podium, thus mimicking a POS situation. Financial incentives were given as a function of performance. Error rates obtained show false reject (Type 1) = 0.2%, false accepts (Type 2) = 0.6%.

Storage required for the references is non-trivial; six signatures are stored, three analogue signals from each for 12.5 s: = 6 × 3 × 1000 bytes, plus the individual's parameters plus other system overheads unspecified say 19 K bytes altogether. Later information indicates that this requirement has been reduced to about 1 Kbyte. This amount of storage is unlikely to be an embarrassment with on line systems or those using an IC card as a portable database, but would frustrate a magnetic card-based system.

De Bruyne at ETH has been active for some years in the development of signature verification methods using a modified form of waveform matching, in which only salient features are correlated. He has also developed a novel 'acoustic radar' digitiser [54].

The method of signature analysis used depends upon the (absolute) timing of several series of similar events. Events used include:

(1) Pen-down pen-up
(2) Isolated reversal of velocity in vertical direction

(3) Isolated reversal of velocity in horizontal direction

(4) Coincidental reversal in vertical and zero in horizontal velocity

(5) Coincidental reversal in horizontal and zero in vertical velocity

Events (4) and (5) correspond to cusps for example.

Time intervals of events from 5 ms to 20 s are recorded on a logarithmic scale of increment approximately 9% giving a 99 point scale. References are scored as mean and standard deviation on this log scale (9 increments = ±2.3 SD range).

The correlation techniques used are based on the comparison of the string of timed events obtained; clearly simple correlation is likely to be unreliable due to the missed or extraneous events. 'Losable' segments i.e. those not appearing in all samples collected at enrolment, are flagged by setting their deviate code to 0. Stored references are updated as signatures are verified and thus changes in user behaviour accommodated.

This technique is much more economical than the IBM approach; though no large scale trials have been undertaken.

*Holistic approach*    A substantial exercise was mounted in the early 1970s at the National Physical Laboratory to explore techniques for signature verification following concern in the UK Inter-Bank Research Organisation for the vulnerability of cheque and credit card operations which were then beginning to be used extensively [55].

The work at NPL was completed in about eighteen months and included collection of a database of signatures (at IBRO offices) and a trial in the entrance hall of a building at NPL. The results of this work were taken up and exploited by Quest Automation [56], and Transaction Security Ltd (now owned by Analytical Instruments, Cambs).

The digitiser used at NPL was pressure sensitive and hence independent of the writing stylus used.

The work at NPL was based on (quasi)-holistic methods in which an arbitrarily fixed number (10) of overall statistical feature values were obtained from a signature.

There are four independent writing 'dimensions' of a signature written on a digitising platten as used in the NPL work, these are:

(1) time

(2) x instantaneous position

(3) y instantaneous position

(4) pen-up pen-down.

From these four properties a range of simple properties or features can be derived;

(5) time to write signature (pen-up for > 1.5 s taken to indicate action finished)₁

(6) total distance travelled (also x and y components)

(7) pen velocity (also independent x and y velocities)
(8) acceleration (also x and y accelerations)

In turn higher order and 'combined' metrics can be derived:

(9) aspect ratio
(10) sum of distances to pen-downs (also pen-ups)
(11) sum of distances to turning points in x (also in y)
(12) sum of time to pen-downs (also pen-ups)
(13) sum of time to turning points in x (also in y)
(14) number of segments
(15) sum of distances (and times) to peaks in acceleration in x (and y).

These features are to a large degree independent. Normalisation by time and/or spatial dimension increases their independence and largely eliminates the effects of those variables.

The original NPL experiments adopted a group of ten measures from the much larger set which can be devised as the basis for experiment. The most useful were ranked on an information theoretic criterion and using this ranking as a basis, the best set of features was selected empirically. The reverse process of selectively discarding the least beneficial ones was too slow with the computing resources available. This early work was undertaken with only 16 K words of RAM and 1.5 M words of magnetic disc; selection procedures can now be much improved and more formally based.

Subsequent commercial products (Quest, Signify, AITSL produced in the early 1980s) have maintained the original philosophy and to a large extent the feature set outlined above, but Signify and TSL have used electromagnetic digitisers which derive some information on pen movement when out of contact with the writing surface and have increased the number of features used.

The potential advantage of allowing measurement of the motion of the pen while out of contact with the surface incurs the penalty of an attached and hence vulnerable pen, a situation strenuously avoided by NPL. Such pens have included a mechanical switch mechanism to detect a pen-down condition.

For experimental purposes a database of signatures was collected independently of the experimenters; this data base contained 3483 signatures from 41 individuals over a period of 13 weeks early in 1973. This data was viewed and cleaned by NPL using a semiautomatic process which discarded 21 maverick signatures (i.e. those with major defects). This material was used exclusively for system design purposes; results quoted below are based on subsequent field trials.

The process of enrolling users is similar to all other systems; the NPL experiment collected an initial five samples which were immediately examined for the presence of any 'wild' samples which were replaced by further samples until a satisfactory 'compact' set was obtained or a total of ten samples still failed to meet the criterion whereupon the user was recorded as unreliable and the set obtained whittled down to five by successively discarding the 'worst' specimens.

Some of the commercial versions seek only two initial signatures for enrolment, depending on subsequent use to improve the user statistics. This stratagem removes the 'session effect' (collection of several signatures at one time exhibited a nontypical consistency between them) but leaves the system vulnerable whilst a better estimate of user signatures is obtained (after several uses of the system).

In use, comparison of the stored reference feature values (mean and SD) with an input specimen were used to derive an error index. A sufficiently low index was accepted (possibly weighted to reflect the individual user's behaviour relative to the behaviour of the user population). Failure to be accepted after two retries resulted in refusal. Individual user statistics were updated on achieving a 'good' match. The criterion for updating was more severe than that for acceptance so that marginally acceptable behaviour was not used to update the measure values. The 'session effect' was noted at NPL and initial references statistics were modified to reduce its effect by combining statistics derived from the whole user population with the individual's statistics, thus weakening the consistency evident in a repetitive signing session and also reducing the effect of novelty at the user's first encounter with the device.

A number of extended trials have been undertaken by the commercial producers, including some in the hands of potential users. These have generally shown equal Type 1 and Type 2 error rates better than 2%.

The NPL work was the first thorough exploration of dynamic signature verification techniques. The author was manager for the NPL project Verisign and also for its exploitation by Quest. There is much more that could be said about the design of experimental situations and the interpretation of results and particularly the problems of comparing results quoted from different sources and situations. The paper by Fox (also part of the Quest team [56]) is illuminating on this topic.

*SRI approach*
The SRI approach has also been developed over a decade and is currently offered commercially by Confirma Inc. It uses an instrumented pen which is constructed to encourage the writer to maintain the tip vertical. The instrumentation consists of three strain gauges coupled to the three orthogonal axes of the writing tip corresponding to nominal vertical (normal to writing surface) and 'x' and 'y' nominal directions in the (writing) plane normal to the tip axis. The appearance of the pen is similar to the familiar UNO STENCIL pen which has a writing head containing an ink reservoir attached to an inclined handle.

Output from the SRI pen is three signals P(t), X(t), and Y(t) representing the forces determined by the three strain gauges.

SRI defined an initial set of 44 features which are derived from these three signals. These include:

Scaled mean
Standard deviation
Minimum
Maximum
Average absolute
Average positive
Number of positive samples
Average negative
Number of negative samples
Number of zero crossings
Maximum minus scaled mean
Scaled mean minus minimum

Also overall features

Total time
Number of segments
Time in contact (pressure $> 0$)

No use of higher order features such as relative timing of discrete events, e.g. pen-up or pen-down, zero-crossing in x or y, is reported.

Discarding formal statistical methods SRI devised an empirical approach for feature selection, based on discarding iteratively the least effective features in a group. As useless or redundant features are removed the equal error rate (T1=T2) decreases to a minimum after which further removals cause the error rate to rise again. The feature set that produces this minimum error rate is taken as the best set. This procedure is an approximation but effective and efficient.

Derived features from a specimen are compared against stored mean and standard deviation values and a simple Euclidean distance computed between the specimen and a reference.

Five to ten true signatures are used to establish initial values for user behaviour statistics at enrolment. If the feature set is reduced to say ten then two numbers must be stored for each feature (mean and standard deviation) for each user i.e. 20 numbers, say 25 bytes.

As described above a common feature set is derived for all users, but it is possible to undertake the feature selection procedure individually for each individual user, thus generating a personalised set. While 10—12 specimen enrolment signatures are adequate for establishing the user values for a standard feature set many more are required to establish personal sets. Starting with 44 features at least 100 specimens should be used. It is possible that personalised sets could be built up during the course of several transactions with the device. The use of personalised feature sets reduced the equal error rate for some individuals from about 6% to essentially zero. (This limited experiment may not be representative).

SRI experiments reported in 1983 used a database containing 5220 signatures obtained from 58 subjects (approximately equal numbers of men and women and including 10% left handers) over a four month period. At each data collection session three signatures were obtained while sitting and a further three while standing. All subjects were told how the system worked and asked to use a consistent signature form. Obvious mistakes were removed from the database. 648 attempts at forgery were collected from 12 'trained' forgers who were told explicitly how the system worked and were allowed video recordings of their targets as they signed. Prizes of up to $100 were available for the best forgeries and most consistent signers.

Results quoted appear to be based on the use of this database in a simulation situation. The following results are given by SRI:

| | Type 1. (true *v*. true) | Type 2 forged *v*. true) |
|---|---|---|
| Error rate | | |
| standing | 1.5% | 2.25% |
| sitting | 1.6% | 3.0% |

As is common experience in this type of situation the majority of the false rejections (Type 1 errors) occurred for a small number of users. If subjects were excluded on the basis of largest standard deviations in their sample set then equal error rate fell to 1.75% with three users excluded and to 0.7% if seven were excluded (from 58 total).

Fuller details of this work are given in the appendices.

### 7.3.3  Gait and grasp

SRI are working on a system for 'gait verification' in which "the user simply walks over an instrumented surface that records the dynamics of the gait". The verification process is based upon comparing various (unspecified) measures of these dynamics with a suitable computer-stored reference. A research prototype has been developed at SRI and evaluated (1982) but does not appear to have been reported.

### 7.3.4  Keyboard rhythms

The potential for identifying (verifying) individual keyboard users by their 'keyboarding characteristics' is reported in some detail by Umphress and Williams in 1985 [57]. SRI also worked on this technique, but no information has been forthcoming from that source. It is understood that the system from International Bioaccess Systems Corp., Ca. is using their technology.

The Umphress and Williams' work follows earlier use of keystroke characteristics in the evaluation of interactive software systems. They characterise the typing task by a series of models. At the lowest level this reduces to

$$Te = \sum_i (Tm_i + Tk_i)$$

where
Te = the total time to execute i keystrokes,
$Tm_i$ = the 'mental time' to select the ith keystroke, and
$Tk_i$ = the time to operate the ith key.

In their experiments they collect statistics, mean and SD, of Te, and the i individual time latencies between keystrokes, $(Tm_i + Tk_i)$, as follows:

Individual users were enrolled by typing a passage of prose containing 1,400 characters and, at a separate time another prose passage of 300 characters.

The initial passage, subject to constraints detailed below, was used to determine a matrix of character pair (digraph) mean latency and standard deviation; also the mean latency and standard deviation for the whole passage.

The second, i.e. test, passage was processed in real-time and the number of times the latency between a pair of character strokes was less than half a SD from the stored mean counted. The mean and SD for the whole test passage was also determined, together with the number of usable keystrokes. Comparison of the overall statistics was based on a standard two-tail t-test, computed and tested as follows:

$$abs (z) > Z_{5\%}$$

where

$Z_{5\%}$ = bounds containing central 5% of the area under a normal curve
and
z    = (testmean-refmean)/(testSD/((testsize)$^{1/2}$))
i.e. the estimate of Student's t.

More than 60% of the character pair latencies have to be within these limits.

Any word containing a spelling or other mistake, or a latency exceeding 0.75 s was discarded, and only the first six characters of any character group were accepted and the remainder discarded.

The values of critical parameters used above were determined empirically.

From a population of 17 programmers, including facile touch typists and others with no formal training the Type 1 error rate was 12%, and Type 2 6%. Unsurprisingly touch typists produced the best results.

Umphress and Williams are at some pains to avoid over optimistic claims for the use of keyboard rhythms alone and suggest that it would be a useful technique in combination with other methods. This is a technique which can be completely covert, requiring only that a user log in in his normal manner. Also that a user can be continuously monitored and any change of operator detected, more or less instantly. This is a major advantage and the only system currently offering such a benefit.


## 7.4  PERFORMANCE

In considering applications of API one of the prime determinants of its effectiveness is the type of user that will be expected to use the system, their motivation, and the level of control that the system management has over the users. In general the use of API systems is to provide the user with a positive benefit, namely to achieve some desired objective, be it entry, money or information. We can assume an initial positive motivation in the user though this may be moderated by apprehension of an unfamiliar 'gadget', and, if the system is unfriendly, a degree of antagonism due to frustration.

The user population can usefully be divided into two distinct classes. Namely:

(1) The open population, and
(2) the closed population.

The open population is essentially the general public which presents itself largely as customers seeking a service from a public service provider; the bank or Post Office are classic examples of such. They are characterised by the large size of their customer base, the wide variety of states amongst their customers, their age, education, tolerance level and so on. Above all the mechanics of the offered service must not offend the customer and must at all times be friendly and helpful.

The closed system, by contrast, is characterised by a firm's employees, and is generally a much smaller population (perhaps tens to hundreds rather than millions). Employees are more inclined to behave according to the firm's instructions; they already accept the requirement to do so in employment. Also they can be penalised for failure to behave in a required manner.

Clearly the performance of a given system is critically dependent on the attitude of its users and no apology is made for labouring the point a little here.

### 7.4.1. Open populations

This is the ultimate application population for any mechanism − exposure to the general public. No API device yet produced is up to this arena. It is essential that any device offer ready use with a minimum chance of a genuine user being refused (and thus 'insulted'). Conversely the protection against fraud must be of a reasonable, but not necessarily very high order, provided that the detection of a potential fraud is reacted to quickly by e.g. a bank operative offering 'assistance', or at least a photograph being automatically taken of the 'user'. If a Type 2 error rate of only 5% is achievable then this represents a 19:1 chance against a fraud being successful and a substantial chance (odds on) of a fraudster being identified and apprehended. This is probably adequate for most practical purposes, though bankers will take some persuading.

A potential weakness in such a system attaches to the variability of a few individual users in their behaviour. It follows that an inconsistent user is more vulnerable to fraudulent attack than are consistent users. There are two acceptable answers to this situation:

(1) Require the inconsistent user to perform additional tests to establish verification, possibly involving assistance from a central operator if the value of the transaction is significant. The use of an IC card goes some way in this direction, and/or
(2) limit the amount of value which said user can obtain.

It is assumed that banking services will be disinclined to refuse facilities to an applicant customer on the basis of inconsistent behaviour, though they have that option. IBM have commented upon the need for all users of a system to get comparable levels of service.

The above comments are related to protection of financial resources. Applications such as time and attendance recording at works entrances need rapid response to cope with mass flow of employees. This requirement precludes many techniques needing a relatively complex protocol to be observed. The use of hand geometry seems appropriate in this situation as it can be rapid and calls for no performance of writing or speaking.

It follows from these remarks that such systems would be introduced within protective premises rather than as required in the longer term in support of unattended terminals in public places.

Performance requirements in this bracket are characterised by a very small Type 1 error, coupled with a modest Type 2 rate say Type 1 = 0.1% and Type 2 = 5%.

### 7.4.2 Closed populations

This is the group of 'professional' users of access to controlled facilities. It is typified by the 'back office' staff in a bank who handle and authorise transfer of large funds, or computer facility staff whose access to sensitive parts of a DP operating system has to be protected and logged to prevent unauthorised interference with the system or access to protected information.

Such personnel are usually already sensitive to the need for security and can be expected (required) to use API systems in a responsible and consistent way. They can also be required to use techniques which are inappropriate with the open population. Such techniques would include retinal scanning, fingerprint verification, keyboard monitoring and speaker recognition under the direction of the system. Speed of operation is likely to be very important but a relatively high rejection rate calling for an independent clearance will be accepted as a concomitant of security.

False acceptance rate Type 2 must be very low, while false rejection Type 1 error may be tolerable at a relatively high level; say Type 1 = 2% and Type 2 = 0.01%. Only EyeDentify and emerging fingerprint systems claim to be substantially better than this though they leave no auditable trail, such as written signatures.

### 7.4.3 Performance assessments

At least two major comparative performance exercises have been mounted in the USA under contract to USAF at MITRE Corp. in 1976 and Sandia Laboratories in 1985. So far the Sandia results have not been openly published; they have been quoted in Personal Indentification News (Washington) though without details of the experiments mounted. MITRE's early results are fully quoted in the literature [58].

As all trials are representative only of the situation in which they are mounted it is most important to exercise caution in comparing trials.

### 7.5  INSTRUMENTATION

In the above description of API devices it is clear that a major effort has gone into the design of input transducer devices. These must be at once cheap (for use in large quantity), familiar to the user, and above all robust in the field. Less evidence is seen of the processing power needed to implement the various algorithms. This must be limited subject to the process time being short, in psychological time, ideally less than a second or so. While on line systems can call upon a large computing resource this will get overstretched in a situation in which a large number of users will present simultaneously but still expect instant service.

Under these circumstances it may be desirable to use the IC card as the processing engine as well as the carrier for the user's database. Such a consideration imposes further limitations on the input transducer to be used which might now be incorporated into the card as well. While it may be possible to include a writing surface, it is unlikely that optical elements can be so the inclusion of a microphone (e.g. electret) is probably the most plausible. An electret element contains little more than an electro-statically charged membrane capacitor and is compatible with the plastic lamination construction likely for IC (or other) cards.


## 7.6 CURRENT R AND D ACTIVITY

It is no surprise that the major efforts in this field, judged by the volume of literature in evidence, are to be found in the USA with major firms such as IBM, Rockwell, TRW, Sylvania and Westinghouse much in evidence over the years, plus a plethora of small entrepreneurial oppor-tunist companies, many associated with or spun off from the larger concerns. Activity is also evident in UK, Japan, Canada, Israel, France, West Germany and Scandinavia. Some European firms have bought in by acquiring US technology − De La Rue Printrak exploits Rockwell Inter-national technology and Siemens have bought out Threshold Technology Inc. Over a hundred firms, institutions and government agencies worldwide have been identified as having or having had substantial activity in the topic of API.

It is evident that much of the innovative work was done in the 1970s. Since that time good products have appeared and much effort has gone into marketing. This effort is now (at last) beginning to bear fruit. New products are again being brought to the market place though apart from the typing rhythm approach, there are no totally novel methods in evidence. Many of the larger firms have technology developed and await the market requirement. This is perhaps a circular situation for it is also evident that the entry of a major company product will legitimise the use of API and hence produce its own market; there is no shortage of concern for the security against intrusion of more or less public transaction and access control systems.

Brief details of some R and D activity past and present can be found in the appendices to this paper.

## 7.7  CONCLUSIONS

From the foregoing it is evident that there is a wide range of API technology which might be adapted to extend the security obtained on the communications and message authentication aspects of the use of ic cards to include also a more thorough authentication of the person presenting such a card. There are a number of major companies who have developed the technology over the past 15 years or so with the clout to introduce complete secure systems, there are also a number of small new companies with novel devices which can trade at the OEM or other licence arrangements with the system implementors. There is growing evidence that, at the third iteration, the market is now convinced of the need for automatic personal identification in support of electronic security systems.

## 7.8  APPENDICES

### 7.8.1  Outline of selected API techniques. Speech based

*7.8.1.1  Texas Instruments*
TI's primary objective was that of physical access control. Their work has been exploited since late 1985 by VoxTron Systems Inc., New Braunfels, Texas for that market.

For verification the user is required to speak a phrase of four monosyllabic words with a consonant-vowel-consonant structure. The audio signal is analysed by a bank of 14 filters over the band 300–3,000 Hz. The rectified output of the filters after smoothing is sampled at 10 ms intervals and digitised into 7 bits in a dynamic range determined by the total energy of the utterance.

Sample utterances collected during an enrolment phase are time aligned on a 'key point' corresponding to the region of maximum energy in the vowel portion of each word. The positions of these key points are recorded for use in subsequent verification procedures. The reference for each word is reduced to a sequence of six spectral amplitude samples taken over a 100 ms period centred on the key point.

In use a four word phrase spoken by the user is split into the four component words and the key point in each located. The reference six spectral samples are then slid along the input sample and the minimum error between them detected. Decision is based on the simple sum of these errors.

Two laboratory trials and a field trial in a TI computer facility produced the following details and results:

(1) The initial trials used a population of 10 female and 40 male users who provided 100 recordings of five phrases over a two month interval. Another group of 70 people providing 20 sessions each were used to represent impostors.

   The samples from the true users were divided into two sets for design and test purposes.

   Data samples were collected under system control but the data was used in purely off line evaluations. Type 2 error performance was derived by using the utterances in the database spoken by notional impostors.

   In the first laboratory experiments an equal error rate of 1.6% was obtained for a single phrase plus a reserve for use if the first phrase failed. Failure to correctly locate the key point was less than 1% for the true user but rose to about 50% for impostors.

   When the test was extended to require two or more phrases to be verified the error rate fell to 0.42% and 0.23% for two and three phrases respectively.

   The number of phrases required to achieve an average Type 1 error rate of 1% was 1.5, with a corresponding Type 2 of 0.01%

   The performance of this initial system degraded with time, indicating that the user's behaviour changed as he became familiar with the system.

   Reported 'non-normal' voices, e.g. with upper respiratory infection, caused a doubling of the Type 1 error rate for affected users.

(2) In the second evaluation the user population was 63 male speakers who attended 25 verification sessions over a month. The impostors were 60 males each attending two sessions. The sample utterances consisted of 16 monosyllabic words organised into 32 four word nonsense phrases.

   At enrolment each of the 16 words was spoken five times. This initial training was extended by four subsequent sessions. These were used to establish estimates of user variability from session to session and moderated weighting factors in the total error measurement. Values obtained from subsequent successful verification were used to update the user's reference. To this extent the system is then adaptive. Verification required four test phrases.

   In this evaluation the single phrase equal error rate was 4%. This improved to 1.5% with two phrases and 0.5% for all four. On average 1.3 phrases were required for Type 1 = 0.3% and Type 2 = 1%.

   It was found that the use of adaptive updating of references progressively reduced the error rates as further utterances were given.

(3) A field trial was mounted at a TI computer facility to control access. Of 180 users, 23 female, an average of 400 entries were made each

day. A four phrase sequential decision strategy achieved a 0.3% and 1% Type 1 and Type 2 error rate respectively.

The average number of utterances was 1.3 per admission, with an average verification time of 5.8 s. To maintain this performance TI required a signal/noise ratio of at least 30 dB.

These results were reported in the late 1970s; it is probable that comparable results are now being achieved in the commercial developments derived from this work. The need for a high S/N ratio is worrying.

### 7.8.1.2 Bell Laboratories

Bell Laboratories began work on speaker verification about 1970. Some members of the team joined TI to perform the work described above. Bell's main interest is the protection of phone channels. The approach adopted chose to employ 'prescribed texts' as test utterances rather than to attempt to detect specific speech events assumed to occur in a text independent approach.

The Bell approach operates in the time domain, based on extraction of 'contours' from the speech signal. These contours correspond to the time function of:

(1) pitch period
(2) gain (intensity)
(3) fourth and eighth coefficients of an eighth order linear predictor (in earlier experiments tracked formant frequencies were used and were replaced with LPC as detection of pitch and 'formants' is difficult with a filter bank).

A sentence-long utterance is used which is sampled at 10 kHz rate. Period and intensity are determined at 10 ms intervals and the formants or LPCs output at 20 ms. The intensity function is used to delimit the beginning and end of the utterance. The four time functions are low-pass filtered and the intensity function is normalised to peak value.

Bell claim that this analysis relates closely to physiological characteristics of inflection and prosody (metre) being "relatively slowly time varying, phase insensitive and broad-band".

Reference utterances collected at enrolment are combined after time registration (using intensity contour and dynamic programming methods) and are length standardised. Each contour is reduced to 20 equispaced samples for storage as a sequence of 80 means and variances after time registration and length standardisation.

The comparison process consists of accumulating the (squared, weighted) difference between reference and sample and a measure of the amount of warp needed to minimise error. The error distance may be either the simple sum of distances at each sample point or a set of points selected separately for each individual.

Four evaluation experiments and a field trial have been undertaken:

(1) Using 40 male speakers and high quality recordings, eight as real customers giving 15 utterances at two day intervals the remainder as impostors giving one utterance. Using three different sets of ten utterances for references and the remaining five for test, an average equal error rate of 1.5% was obtained. This increased to 6% if only end point alignment was used.

In this evaluation formants were used as contours to be replaced in later experiments by LPCs. The formant (or LPCs) were found to be of little value in confirming the true user.

(2) Introducing attack from four trained mimics Type 2 error rose to 27%. It was evident that the formant or LPCs were very important here under this form of attack.

(3) Human listeners to attempt recognition of paired utterances as the same speaker or not. These tests attained an equal error rate of 4%; with mimics included Type 2 increased to 22%. Clearly the system is better at straightforward test and significantly worse with mimics included.

(4) With another database of 22 male users giving 50 recordings over a two month period, and 50 impostors giving one utterance each; then, using fourth and eighth prediction coefficients:

(a) using speaker dependent contours references, equal error = 1.5%; this rises to 4% using speaker independent reference methods. The Type 2 error rate with the mimics was 4%.

(b) if only pitch period and intensity are used, equal error rates rise to 3%, 6%, and 16% respectively.

(5) In a limited 'real-world' trial a 100 'customers' used the system over dial-up lines within Bell; five samples collected in a single session were used as references, daily use was required, sessions were controlled by spoken instruction and users claimed identity by numeric code. References were updated as described above, each reference required 1200 bytes per customer. Only the pitch-period and intensity profiles were used.

The equal error rate was 9–10% over a five month trial period during which customers managed on average some 50 sessions each; some 4500 verification sessions in total. This performance had improved to 4% after 20 sessions per user.

Clearly the vulnerability to mimicry is very worrying, but results are not markedly different from other techniques.

*7.8.1.3   Threshold Technology*
Experiments were reported in 1981, by Threshold Technology, using much of their standard speech recognition hardware as a front-end speech

processor. They comment that typical phone system distortion does not encourage the use of techniques based on spectral power or zerocrossings.

Instead they performed formant tracking on the output from a filter bank of 16 channels in the band 260 Hz to 4.5 kHz. 32 binary features are extracted from this data every 2.2 ms; these features consist of ten energy maxima (formants) in the range 260–1892 Hz, 18 energy slopes in the range 260–1405 Hz, and four 'miscellaneous' features. Phoneme-like features used in the recognition product were discarded as too expensive for the purpose. Word boundaries were detected and the utterance (of single words) divided into 16 equal time slots.

In the enrolment procedure a repertoire of 16 selected words were spoken five or ten times. An average template is constructed whilst using non-linear time warping (using dynamic programming) to obtain the best fit between them.

For verification the processed sample utterance is matched against the corresponding reference, again using time warping, to determine the sum of the individual errors in matching a number of words.

Published results are rather sparse and preliminary.

Tests used 42 speakers (27 male), trained in one session and verified on ten subsequent days. Training was undertaken in two ways, sequentially when a list of words is read through five times; and repetitively when the same words are individually repeated five times.

The performance was analysed for male and female customers.

|  |  | Equal error rate % |
|---|---|---|
| Sequential training | All speakers | 1.5 |
|  | Males | 1.3 |
|  | Female | 2.4 |
| Repetitive training | All speakers | 1.4 |
|  | Male | 0.9 |
|  | Female | 2.2 |

The relatively poorer performance for females is attributed to the wider separation of spectral lines due to the higher pitch of female voices, this results in poorer definition of the formant structure.

This approach is akin to the (earlier) Bell work. Threshold Technology have now become part of the Siemens empire and they are marketing a system SIPASS which uses speaker verification, presumably based on this technology.

### 7.8.1.4 *Philips*

The first publications by Philips on speaker verification date from 1977. The primary interest for the work has been access control.

Work at Philips prior to 1982 used long-term spectra (LTS) of speaker's prescribed utterances. The use of LTS avoids the need to determine any internal detailed structure of the utterance beyond the approximate start and finish points.

A 43 channel filter bank is used in the band 100 Hz−6200 Hz which is sampled every 20 ms. The filter outputs are integrated over the duration of the utterance. By averaging adjacent channels the number of components is reduced to 15 and their dynamic range represented on a nonlinear (quasi log) scale of 6 bits.

An experimental database was collected from 50 speakers who gave ten 12 second utterances at five sessions with at least a week between each session, 2500 sample utterances in all. The first 20 sessions' utterances were taken as a training set and the remainder as a test set in experiments detailed below.

References were generated by determining the mean amplitude and variance of the 15 components from the design set of utterances. A clustering test was applied to the reference utterances in order to discard 'outliers' i.e. utterances which for some reason were not characteristic (due for example to incomplete utterances for whatever reason).

Test specimens were compared against the references by both a simple threshold on the summed weighted difference between the individual pairs of corresponding averaged spectral components, and a more complex criterion using a nearest neighbour distance criterion on each of the reference utterances (thus defining a piecewise linear bounded volume). The latter is a more powerful classifier as it discriminates on the errors occurring with each of the individual references rather than on the error against the mean (weighted by the variance) for the pooled references.

The nearest neighbour metric requires approx n/2 times as much storage as the simpler minimum distance approach, where n is the number of utterances in the reference set.

Experimental results reported are limited to the statement that for the minimum distance classifier an equal error rate was around 0.9%. This increased markedly if the resolution of the stored data was reduced, while increasing it had no significant effect. The storage required was 188 bits per reference. The best result with the nearest neighbour approach gave an equal error rate of 0.8% requiring 1298 bits per reference.

Later work using a Bayesian classifier and methods of data processing and storage in logarithmic form (thus avoiding the need for multiplication in the associated arithmetic) used the same processing method but with the sampling interval increased to 27 ms. References were updated as utterances were accepted, thus improving the representativeness of the references as more samples became available. Up to three tries were allowed to gain acceptance.

Performance is then said to better the USAF (BISS) requirement of

Type 1 error rate <1% and Type 2 error <2%. References were allowed 256 bytes per user.

Later work by Philips characterises speech utterances by the contours of intensity and pitch in processes very similar to Bell Laboratories. Philips have, however, investigated more sophisticated weighting procedures for use in the comparison phase. These include effects of both amplitude and temporal variation by the user and aspects of the total user population variation for each feature (corresponding to short-term spectra). By including aspects of the total population's behaviour it is shown that a two-fold improvement in performance is achievable.

The additional storage requirement for nearest neighbour comparison techniques is only relevant if storage is at a premium (e.g. on a mag-stripe card). The smart card will not be so restrictive.

### 7.8.1.5  *Carnegie-Mellon University*
Work by Shikano presented in 1985 and other earlier workers describes recent work aimed at developing a speaker and context independent speech recognition system in which as a first requirement the user, from a closed population of users, is first identified in order to select the appropriate set of speech 'templates' for that speaker. A little introspection will reveal that as facile speech recognisers we adjust our perception of a person's utterance to accommodate idiosyncratic properties of that speech e.g. lisp, regional accent, even expected vocabulary, so this approach for mechanical speech recogniser is not without precedent.

This work used a population of nine male speakers who each spoke a set of ten sentences of approximately 2.5 s duration (the Harvard Sentences). These signals were stored digitally after sampling at 16 kHz. Of these ten sentences five were used for analysis purposes the remaining five for testing, thus ensuring context freedom.

Processing consisted of LPC analysis to yield up to 256 vectors for the utterance. Using the full set of vectors and the complete utterance, a recognition rate of 98% was attained. Less detailed representation and/or truncated utterances produced inferior results.

Use of 64 vectors on the full utterances or 128 vectors on half the length reduced performance to 93%. See paper for much more detail.

### 7.8.2  Methods using signatures

### 7.8.2.1  *IBM approach*
The remarks below describe a well documented experimental situation, other work by Lew indicates further benefit to be gained from a more thorough understanding of the kinematics of writing (but requiring more instrumentation of the pen to fully determine all six dimensions of motion).

The IBM approach uses an instrumented pen which contains two accelerometers and a tip-pressure transducer. The accelerometers are mutually orthogonal and measure acceleration normal to the axis of the pen. The pen is of cylindrical form thus making no demands on the user with respect to orientation. As the attitude of the accelerometers is thus undefined in use, the signals from them are combined during a (software) signal conditioning process, to indicate acceleration with rotational invariance but retaining sign. The pressure transducer like the accelerometers is piezo-electric and gives an analogue output, and because of its rigid structure imparts no additional 'feel' to the user (as is the case with mechanical switch mechanisms to detect pen-down).

After signal conditioning which includes band limiting filtering, the analogue signals are sampled at 80 Hz into eight bits per sample per signal channel. The pressure signal is the first differential of pressure due to the nature of the transducer (piezo-electric) and associated amplifier.

In the most recent (and detailed) description of the IBM approach the algorithms used for comparing signals from a given signature with those from a nominated target for verification are presented. All processing is on the basis waveform comparison, taking each segment individually, with adjustments for very long duration segments.

Through examination of the pressure (derivative) signal, segments are separated and used as a basis for all further processing and comparison. Segments extending for more than 0.7 s are artificially segmented.

Assessment of the similarity of an offered signature to a reference is based on a set of five measures:

M1 Segment alignment
M2 Pressure correlation
M3 Acceleration correlation
M4 Pressure coherence
M5 Acceleration coherence.

These are defined as follows:

$$\text{M1} = \frac{1}{\text{NSEGS}} \sum_{i=1}^{\text{NSEGS}} \frac{(\text{seglr}_i - \text{segls}_i)^2}{\text{seglr}_i \times \text{segls}_i}$$

where $\text{seglr}_i$, $\text{segls}_i$ are the segment lengths (duration) of segment i of reference and specimen respectively.

$$\text{M2} = \sum_{i=1}^{\text{NSEGS}} \text{olapp}_i \times \text{pscore}_i$$

where $\text{olapp}_i$ is a weighting factor computed as the amount of time the two segments overlap divided by the total pen-down time of the reference; and

$$\text{pscore}_i = \frac{\text{MAX}}{\text{tau}} \left\{ \frac{\sum_t S_p^i (t) R_p^i (t + \text{tau})}{\{\sum_t S_p^i (t)^2\}^{\frac{1}{2}} \times \{\sum_t R_p^i (t)^2\}^{\frac{1}{2}}} \right\}$$

$\text{pscore}_i$ is the maximum value found over an allowed time lag tau. This is dependent upon the initial alignment position of the segment.

$$M3 = \sum_{i=1}^{\text{NSEGS}} \text{olapa}_i \times \text{ascore}_i$$

where $\text{olapa}_i$ and $\text{ascore}_i$ represent acceleration and are defined in a way analogous to $\text{olapp}_i$ and $\text{pscore}_i$ respectively.

$$M4 = \sum_{n=1}^{5} w_p(nf_0) \times z_p(nf_0)$$

where $z_p(nf)$ represents the relative power contained in the signals at frequency f and is evaluated as

$$z_p(f) = \left\{ \frac{\text{MOD}(G_{rs}(f))^2}{\text{MOD}(G_{rr}(f)) \times \text{MOD}(G_{ss}(f))} \right\}^{\frac{1}{2}}$$

and $w_p(nf_0)$ is a suitable weighting function.
Grs(f), Grr(f) and Gss(f) are the cross- and auto-spectral density functions of the reference and specimen pressure signals respectively, evaluated as:

$$\frac{1}{\text{NSEGS}} \sum_{i=1}^{\text{NSEGS}} Sf(f)_i \times Rf(f)_i$$

$Sf(f)_i$, $Rf(f)_i$ are the Fourier coefficients of the specimen and reference pseudo-signature's ith segment. Frequency range 2.5–40 Hz.

$$M5 = \sum_{n=2}^{5} w_a(nf_0) \times z_a(nf_0)$$

which is analogous to the pressure coherence measure but the frequency range is from 5–12.5 Hz.

In determining M1 it is necessary that the number of segments are the same in specimen and reference; this condition is not likely to be met reliably from one signing to another. Accordingly individual segments in the specimen may be merged (adjacently) or divided in such a way that M1 is minimised, before any long segments are split up.

A Similarity Measure, SM, between a specimen and a reference signature pair is computed as:

$$SM = \sum_i Sm_i$$

where $Sm_i$ is calculated as following:

$$Sm_i = \frac{(M_i - MID_i)^3}{(HI_i - MID_i)^3} \qquad M_i >= MID_i$$

$$Sm_i = \frac{(M_i - MID_i)^3}{(MID_i - LO_i)^3} \qquad M_i < MID_i$$

in which $MID_i$, $LO_i$ and $HI_i$ are chosen by a mechanism which reflects the expected values of that measure when applied to a body of genuine and forged signatures.

These formulae yield an index which indicates the degree of match on a monotonically increasing scale. A threshold value applied to this index then adjudicates acceptability or otherwise of the specimen offered against a reference.

User references are established as follows:

A set of N signatures are collected from a user at enrolment, $N > 2$.

From these signatures two are selected as first and second references by computing the SM for each signature against every other signature in the set (of N). Find the minimum of these maximum values. Finally select the two signatures that have the maximum of the found minima.

This technique is used to accommodate users who present two signature styles. If a 'maverick' is detected it is discarded and a new pair selected from the reduced set.

In a practical enrolment situation some six sample specimen signatures are collected. Apart from the first and second references a 'bucket' of four more signatures is maintained which contains the four most recently accepted signatures so that a set of six signatures is always available from which the two references can be selected. This mechanism ensures that the system accommodates gradual changes in the user's behaviour.

The decision mechanism is based on the SM values obtained from the first and then second reference or if either are only marginally rejected then the signatures in the bucket are also used.

In order to increase the protection against successful forgery and to reflect the difference in the degree of variability between users a number of individual parameters are determined at enrolment (and subsequently updated) for each user and are stored with his references. Namely:

QUIET — the length of quiet sections i.e. pen out of contact
MAXPTS — the maximum segment length
SHIFTS — the percentage of shift needed to maximise regional pressure correlation.
$LO_i$, $MID_i$, and $HI_i$ — five sets of similarity component parameters
REF — an overall 'adjustment parameter' for the SM

If a signature is found unacceptable then one or two more samples are requested. If they also fail then the user is rejected.

In discussing the measurement of performance IBM note that all users

should receive a similar level of service and define two further useful
factors to characterise performance in addition to the averaged Type 1
and Type 2 values obtained. These are:

$U(X)$ − the utilisation index = $1/CP(X)$
　　　where $CP(X)$ = the cumulative percentage of users who achieve
　　　$X\%$ of Type 1 errors
$Z$ − the percentage of users with zero Type 1 errors = $CP(0)$

Of 4,625 sessions only nine Type 1 errors occurred and 4,700 signatures
were collected; no user had more than one refusal. Type 2 errors were
based on 2,133 signatures collected in 1,068 sessions. The total number of
forger-target combinations was 592. Six Type 2 errors occurred. In two
cases successful forgeries were obtained in the two sessions in a single day
(it is implied by this remark that the targets were changed on a daily
basis).

In order to investigate the effect of variation in manufacturing tolerances
in the pen, several pens were used and interchanged during field trial
tests.

Error rates obtained show false reject (Type 1) = 0.2%, false accepts
(Type 2) = 0.6%.

The parameter $Z = 91.6\%$, $U(100) = 4.8\%$, $U(99) = 2.8\%$ and $U(95)$
$= 2.0\%$.

The system has been implemented on the VM network (using a remote
PC as a terminal) and on an IBM/PC (standalone).

This work has progressed continuously over at least ten years and is
well documented. The above account is abstracted from the comprehensive
paper by Worthington *et al.* which should be consulted for further detail.

### 7.8.2.2   de Bruyne approach (ETH)

De Bruyne at ETH has been active for some years in the development of
signature verification methods and the concomitant development of a
digitiser using acoustic ranging techniques. This digitiser uses linear strip
microphones on two orthogonal axes to receive an acoustic transient (by
electrical discharge) from a stylus which has a spark gap near the writing
tip and a pressure operated switch to indicate pen-up pen-down. The
spark is initiated by the measurement system so that delay between
triggering the spark and the arrival of the associated event at the microphones indicates range. The electrostatic microphones used can also be
used as transmitter/receivers in a fully active 'acoustic radar' approach.

The method of signature analysis used depends upon the (absolute)
timing of several series of similar events. Events in a signature
include:

(1) Pen-down pen-up
(2) Isolated reversal of velocity in vertical direction
(3) Isolated reversal of velocity in horizontal direction
(4) Coincidental reversal in vertical and zero in horizontal velocity
(5) Coincidental reversal in horizontal and zero in vertical velocity

Events (4) and (5) correspond to cusps for example.

Periods between events from 5 ms to 20 s are exhibited and are recorded on a logarithmic scale of increment approximately 9% giving a 99 point scale. References are scored as mean and standard deviation on this log scale (9 increments = ±2.3 SD range).

No details are given of the correlation techniques used on the string of timed events obtained; clearly simple correlation is likely to be unreliable due to the inevitable missing or inclusion of events. 'Lost' segments (i.e. not appearing in all samples collected at enrolment) detected during the enrolment sequence are flagged by setting the deviate code to 0.

A method is referred for updating stored references as signatures are verified in order to track changes in user behaviour. This technique is much more economical than the IBM approach; so far no large scale trials have been reported.

### 7.8.2.3 SRI approach

SRI use a pen fitted with strain gauges to measure writing forces in three orthogonal directions, namely pressure (P), and the two directions in the plane (X, Y). They defined an initial set of 44 features which are derived from these three signals; these include viz:

| Source signal | | | Feature property |
| X(t) | Y(t) | P(t) | |
| Feature number | | | |
| --- | --- | --- | --- |
| 1 | 11 | 21 | Scaled mean |
| 2 | 12 | 22 | Standard deviation |
| 3 | 13 | 23 | Minimum |
| 4 | 14 | 24 | Maximum |
| 5 | 15 | 25 | Average absolute |
| 6 | 16 | 26 | Average positive |
| 7 | 17 | 27 | Number of positive samples |
| 8 | 18 | 28 | Average negative |
| 9 | 19 | 29 | Number of negative samples |
| 10 | 20 | 30 | Number of zero crossings |
| 31 | 32 | 33 | Maximum minus scaled mean |
| 34 | 35 | 36 | Scaled mean minus minimum |

Also overall features

| | |
|---|---|
| 40 | Total time |
| 41 | Number of segments $- 1$ |
| 42 | Time up $(P = 0)$ |
| 43 | Number of segments |
| 44 | Time down $(P > 0)$ |

Clearly some of these later features are directly related. No attempt to define or use higher order features such as relative timing of discrete events, e.g. pen-up or pen-down, zero-crossing in x or y, is reported.

SRI have tried a number of standard statistical methods to computerise feature selection from the group of 44 for use. These methods have included: univariate F-ratio evaluation, Fisher's discrimination analysis, information measures such as divergence. Although these methods perform reasonably well they are all based on a number of unrealistic assumptions; of Gaussian distribution, equality of the covariance matrices and the like.

SRI devised an approach which resulted in an improved feature selection and lower Type 1/Type 2 error rates. It is summarised as follows:

(1) Assume that the initial feature set contains N features.
(2) Calculate Type 1/Type 2 error curves for all N-1 subsets.
(3) Using the criterion of minimal equal error rate calculate the T1/T2 error curve for all N-2 subsets of the best N-1 subset.

and so on.

As useless or redundant features are removed the equal error rate (T1 = T2) decreases to a minimum after which further removals cause the error rate to rise again. The feature set that produces the minimum error rate is taken as the best set. This procedure is an approximation and more complex processes are possible but require a large amount of computing.

The method described makes no assumption about the underlying probability distribution of the feature set; involves simple and intuitively reasonable calculations; selects a 'best' set of features; and reduces the feature set while converging on a minimum error point.

In computing the distance metric it was assumed that the features are statistically independent (which they are not, but they do not conform to any convenient form of e.g. multivariate Gaussian so no other assumption can reasonably be made either). The distance metric used was the intuitively satisfying and simply calculated weighted Euclidean:

$$d(\bar{S}) = \left\{ \frac{1}{f} \sum_{i=1}^{f} \frac{(S_i - T_i)^2}{(V_i)} \right\}^{\frac{1}{2}}$$

where $\bar{S}$ is the sample feature vector of components Si (i=1 to f) and Ti is the true (reference) feature mean value of variance Vi.

This distance measure requires only five to ten true signatures to establish initial values for user behaviour statistics at enrolment.

The 'best' feature set selected as standard by the above method contained features: 1, 2, 3, 6, 11, 12, 13, 14, 16, 20, 22, 25, 26, 27, 28, 29, 30, 32, 33, 40, 41, 42, 43 and 44. 24 features in all.

## REFERENCES

[1] Masuyama, H. (1985) 'Properties of personal identification systems using question−answer techniques' *Trans Int Electronics and Communications Engineering* J69D/4, 613−620.

[2] Kniessler, H. (1985) *Identification of individuals with computer graphics* SRI Report, SRI International Inc.

[3] Harmon, L. D., Khan, M. K., Lasch, R. and Ramig, P. F. (1981) 'Machine identification of human faces' *Pattern Recognition* 13(2) 97−110.

[4] Siemens, A. G. (1982) *Fingerprint personal identification system* DE 3036 912. 13 May 1982.

[5] Asai Koh (1980) *Device for extracting a density as one of a number of pattern features extracted for each feature point* ... GB 2050026. 31 Dec 1980. *Nippon Electric Co Ltd.*

[6] Sparrow, M. K. and Sparrow, P. J. (1985) *A Topological approach to the Matching of Single Fingerprints* US Dept of Commerce, NBS. Oct. 1985.

[7] Isenor, D. K. and Zaky, S. G. (1986) 'Fingerprint identification using graph matching' *Pattern Recognition* 19(2) 113−122.

[8] Sibany Mfg. Corp. (1966) *Recognising fingerprints* GB 1150511. 29 April 1966.

[9] Stellar Systems Inc. (1983) *Personnel identification system using characteristic data* US Au8335163. 31 March 1983.

[10] Fowler, R. C., Ruby, K., Sartor, F. F. and Sartor, T. F. (1985) *Fingerprint Imaging Apparatus* US 4537484. 27 Aug 1985. Identix Inc, Palo Alto, Ca.

[11] Fingermatrix Inc. (1984) *Fast action fingerprint check for access control* US EP125−532. 21 Nov 1984.

[12] IBM Corp. (1979) *Palm print identifier* GB 1535467.

[13] Palmguard Inc. (1982) *Image recognition system for recognising human palm.* US 4186378. 20 Jan 1982.

[14] Stellar Systems Inc. (1984) *Personnel Identification Devices Using Hand Measurement Techniques* April, 1984.

[15] Mitsubishi Denki K. K. (1985) *Individual identification apparatus based on finger length comparison* JP EP132665A. 13 Feb 1985.

[16] Zuccarelli, H. (1983) 'Ears hear by making sounds' *New Scientist* **100** BPC.

[17] EYE-D Developments II Ltd (1984) *Ocular fundus reflectivity pattern identification apparatus* US EP126549A. 28 Nov 1984.

[18] Rice, J. (1986) *Method of, and Apparatus for the Identification of Individuals* UK 8509389 86.

[19] Abberton, E. and Fourcin, A. J. (1978) 'Intonation and speaker identification' *Language & Speech* **21**(4) 305.

[20] Bolt, R. H., Cooper, F. S., David Jr, E. E. and Denes, P. B. (1973) 'Speaker identification by speech spectrograms: some further observations' *JASA* **54**(2) 531−537. Acoustic Society of America.

[21] Holden, A. D. C. and Cheung, J. Y. (1977) The role of idiosyncracies in linguistic stress cues, and accurate formant analysis, in Speaker Identification *Carnahan Conference 1977*, University of Kentucky, 33−37.

[22] McGlone, R. E. and Hollien, H. (1976) 'Partial analysis of acoustic signal of stressed and unstressed speech' *Carnahan Conference*, University of Kentucky, 19

[23] Davis, R. L., Sinnamon, J. T. and Cox, D. L. (1982) *Voice verification upgrade*. Report Texas Instruments Inc.

[24] Rosenberg, A. E. (1976) 'Automatic speaker verification: a review' *Proc IEEE* **64**(4) 475−487.

[25] DeGeorge, M. (1981) 'Experiments in automatic speaker verification' *Carnahan Conference 1981*, University of Kentucky, 103−110.

[26] Bunge, E., Hofker, U., Hohne, H. D. and Kriener, B. (1977) 'Report about speaker-recognition investigations with the AUROS system' *Frequenz* **31**(12) 382.

[27] Kuhn, M. H. and Geppert, R. (1980) 'A low cost speaker verification device' *Carnahan Conf 1980*, University of Kentucky, 57−61.

[28] Ney, H. and Gierloff, R. (1982) 'Speaker recognition using a weighting technique' *IEEE Int. Conf. on Acoustics, Speech and Signal Processing* 1645−1648

[29] IBM Corp. (1967) *Speech recognition* GB 1179029. 19 April 1967.

[30] Perkin Elmer Corp. (1970) *Speech recognition* GB 1289202. 21 April 1970.

[31] NCR Corp (1977) *Verification* GB 1532944. 21 Feb 1977.

[32] Kashyap, R. L. 'Speaker recognition from an unknown utterance and speaker-speech interaction' *Tr Acou. Speech & Signal Processing ASSP* **24**(6) 451.

[33] Tokyo Shibaura Denki (1983) *Individual verification apparatus based on speech recognition* JP EP-86−064. 27 Jan 1983.

[34] Reitboeck, H. J. (1977). 'Speaker identification over telephone transmission channels' *Carnahan 1977*, University of Kentucky, 237−238.

[35] Shridhar, M., Baraniecki, M. and Mohankrishnan, N. (1982) 'A unified approach to speaker verification with noisy speech input' *Speech Communication* 103−112.

[36] Kiyohiro, S. (1985) *Text-Independent Speaker Recognition Experiments using Codebook* Computer Science Dept C-M U, 9 April 1985.

[37] Moore, R. K. 'Systems for isolated and connected word recognition' *NATO ASI* **16** 74−143 Springer-Verlag.

[38] Nagel, R. N. and Rosenfeld, A. (1977) 'Computer detection of freehand forgeries' *Trans on Computers* C **26**(9) 895.

[39] Rediffusion Computers Ltd (1983) 'Rediffusion introduces low-cost signature verifier' *Financial Times* 24 Feb 1983, 19.

[40] Chainer, T. J., Scranton, R. A. and Worthington T. K. (1985) *Data input pen for signature verification* US 4513437. 23 April 1985.

[41] Crane, H. D. and Ostrem, J. S. (1983) 'Automatic signature verification

using a three-axis force-sensitive pen' *Trans on Sys., Man & Cybernet* **SMC-13**(3) 329–337.

[42] Quest Automation Ltd (1979) *Transducer pad for electrographics* GB 1539755.

[43] Greenaway, D. L. (1978) *Apparatus and Method for Producing an Electrical Signal Responsive to Handwriting Characteristics* US 4122435. 24 Oct 1978.

[44] Darringer, J. A., Denil, N. J. and Evangelisti, C. J. (1975) 'Speed pen' *IBM Tech Disclosure Bulletin* **18**(7) 2374–5.

[45] Radice, P. F. (1980) *Personal Verification Device* US 4234868. 18 Nov 1980.

[46] Worthington, T. K., Chainer T. J., Williford, J. D. and Gundersen, S. C. (1985) 'IBM dynamic signature **verification**' *Computer Security*, Elsevier Science Publishers BV, 129–54.

[47] Zimmermann, K. P. and Varady, M. J. 'Handwriter, identification from one-bit quantized pressure patterns' *Pattern Recognition* **18**[1] 63–72.

[48] Hale, W. J. and Paganini, B. J. (1980) 'An automatic personal verification system based on signature writing habits' *Carnahan Conf. on Crime Counter*, University of Kentucky, 121–125.

[49] Stuckert, P. E. (1979) 'Magnetic pen and tablet' *IBM Technical Disclosure Bul* **22**(3) 1245–1251.

[50] de Bruyne, P. (1984) 'An ultrasonic radar graphic input tablet' *Scienta Electrica* 1–26.

[51] Lew, J. S. (1980). 'Optimal accelerometer layouts for data recovery in signature verification' *IBM Journal of R & D* **24**(4) 496–511.

[52] Herbst, N. M. and Liu, C. N. (1977) 'Automatic signature verification based on accelerometry' *IBM Jour of R & D* 245.

[53] Liu, C. N., Herbst, N. M. and Anthony, N. J. (1979) 'Automatic signature verification: system description and field test results' *TR Sys., Man & Cybernetics* **SNC 9**(1) 35.

[54] de Bruyne, P. (1977) 'Developments in signature verification' *Int Conf on Crime Countermeasure*, Oxford Univ.

[55] Watson, R. S. and Pobgee, P. J. (1980) 'A computer to check signatures' *Visible Language* **13**(3).

[56] Fox, P. F. (1982) 'A practical method of personal identification by signature validation' *IACSS Conference*, Zurich, Switzerland.

[57] Umphress, D. and Williams, G. (1985) 'Identity verification through keyboard characteristics' *Int J Man-Machine Studies* 263–273.

[58] Haberman, W. and Fejfar, A. (1976) 'Automatic identification of personnel through, speaker and signature verification – system design and testing' *Carnahan Conference 1976*, University of Kentucky, 23–30.

Chapter 8

# Cryptography and the Smart Card

## D. W. DAVIES

(Data Security Consultant)

*Cryptology is the key technology for secure systems.*

## 8.1 Introduction

The close relationship between smart card and cryptographic techniques can be looked at from two directions. The smart card can be used as a component of a cryptographic system to improve its convenience or level of security. From the other viewpoint, the smart card itself is the main component of the system and cryptography is called upon to help it with its task. In this chapter we shall mainly adopt the second viewpoint, which is centred on smart card applications but first let us look at the smart card as an adjunct to cryptography.

The confidentiality of data on a communication line can be protected by enciphering it. Encipherment is a transformation which makes the transformed data seem meaningless to an outsider, yet which allows an inverse transformation, for those authorised to receive the information, which turns it back into its clear text. To separate the authorised readers from others, the authorised readers hold a secret value called a *cryptographic key* without which decipherment is impossible. In the usual form of cryptography, this secret key is used as a parameter for both the encipherment and the decipherment functions.

When cryptography is used to protect data travelling some distance, before it can go into operation a secret key must be established at both the sending and the receiving end. Conveying the key from one place to the other entails a risk of losing it to an opponent. A smart card can be used to store a key for secure transport. The use of this key can be authorised by means of a password, known only to authorised users, and the smart card itself can take part in the complex process of key management. Some of the techniques are described later in the chapter.

Sometimes, cryptography is used to encipher information not for communications purposes but to protect it while it is stored locally. It might be difficult to protect the local store from illicit access or information stored on a removable medium might be stolen or copied. When cryptography is used for stored data, the keys are not transported but their security is very important because they can unlock all the protection provided. Most computers are physically insecure, so a smart card can be used to hold the keys and the card taken away by its owner and stored in a safe place. Here also, a password can be used to unlock the secret key from the card.

A related problem of cryptography is the protection of the cryptographic mechanism itself. Not only must the key be protected but also the place where the cryptographic transformations take place. Smart cards can help in this problem by becoming, themselves, the 'cryptographic engine' of the system. If they have enough processing power for the purpose, they can hold all the protective mechanism of a secure system, particularly at the terminal end where the processing demands are less severe. The computer itself, which might be an intelligent terminal, is physically insecure and any part of its store or process is open to tapping or 'bugging'. To counter this, cryptographic methods are used and the keys, together with the cryptographic transformations, are contained entirely in smart cards. When these are removed, the system is locked up and the information it contains is safe against illegal access.

These are examples of the close relationship between smart cards and cryptography, seen from the side of the cryptographer who regards the smart card as an additional tool. Our viewpoint in the rest of this chapter is to think of the smart card as a main component of the system and see how cryptographic techniques are used for its purpose.

## 8.2  PROTECTION FROM PASSIVE AND ACTIVE ATTACKS

Cryptographic techniques can be used in a large number of ways and for many different purposes. The basic purpose is to protect a system against misuse by impostors or unauthorised people. The first stage in protecting a system is to analyse the threats to the system and the risks they entail. We shall consider only those threats that are amenable to cryptographic protection and, as a first step we divide these into *passive* and *active* attacks.

A passive attack attempts to read information without changing it. Examples are the tapping of a telephone line, stealing or copying a diskette, observing a password by looking at the keyboard while it is entered or picking up stray electromagnetic radiation from which a meaningful signal can be reconstructed. Generally speaking, these attacks are not difficult to carry out and in a widespread communication network it is impossible to prevent them. The tapping or bugging of voice conversations is a highly developed art which can be applied (with a few changes) to the collecting of digital data from communication lines, I/O channels, processors, stores, keyboards or any other part of the information system. In these circumstances, to preserve the confidentiality of information it must be transformed by encipherment and then transformed back into clear form for processing, printing or display. Wherever the information is in clear form there must be other ways to protect it, such as not allowing unauthorised persons into offices where information is displayed or printed.

On the other hand, an active attack is one which seeks to alter the information, perhaps to falsify a transaction, prevent a debit to a bank account from reaching it or even destroy an entire file. Generally speaking, these active attacks are much more difficult to carry out and they require more skill and sometimes more luck to achieve their purpose. When they do succeed, the consequences are often more serious than those of a passive attack.

It might seem that encipherment was all that is needed to prevent an active attack on data but this is not so. Some of the more extreme active attacks such as placing a bomb in the computer room are not amenable to cryptographic protection but, these aside, there is a much wider range of possibilities against which protection is needed. Suppose, for example, that an important file is stored in enciphered form and updated from time to time, while still being enciphered. If we are not careful, an attacker simply replaces today's file by an older one which will pass as genuine because it is properly enciphered. Whole transactions might be knocked out on a communications line without detection. Some encipherment methods allow an attacker to change individual bits of a message or file without knowing the cryptographic key. Thus there is more to protection against an active attack than encipherment alone and the methods used have in the past been known as 'authentication' because they ensure that the information remains authentic. This term is discouraged by the recent usage of the International Standards Organisation and we speak therefore of *data integrity* meaning that the data takes the values it was intended, not those altered or substituted by an impostor.

In an extended communication system it may not be possible to prevent the changing of information but we can at least hope to detect when the information loses its integrity. Thus we shall normally be speaking about integrity verification rather than the prevention of an active attack.

One form of active attack is the *masquerade* which means an unauthorised user pretending to the system that he is an authorised one. It goes further than this because any part of the system can be subject to a masquerade. A well known example is the program which asks the user to present his password. If the program is an impostor, the passwords can be given away to an enemy. Protection against masquerading should therefore be included in every interaction within the system where it is physically possible for an intruder to slip in. In the work of the International Standards Organisation this is known as *peer entity authentication* since the two communicating parts of the system are entities at the same level, i.e. *peer entities*. The particular case which we meet most frequently is the authentication of a user to the system. Masquerading as an authorised user is the commonest type of active attack on teleprocessing systems, typified by the 'hacker' who mainly has to contend with rather simple password systems. Authentication of users can take much more secure forms, using smart cards, which will be described later in section 8.5. As in the present use of 'automatic teller machines' (cash dispensers), a token which the user holds together with a password which he remembers adds a lot to the security. A smart card, which is very difficult to forge, makes an ideal token for this purpose.

Masquerading forms a bridge between passive and active attacks. By tapping the line to find the password or guessing the password or trying a lot of passwords the attacker obtains the means to enter the system in the guise of a genuine user. This gives him the possibility of both a passive and an active attack, the active attack requiring no more skill than normal use of the terminal, if a public network is used for access.

In the next three sections of this chapter we shall describe the use of cryptography against passive attacks, preservation of data integrity and user authentication.

## 8.3  CRYPTOGRAPHY

Cryptography is an ancient art which, like many others, is changing rapidly with the development of information processing technology. We are concerned only with ciphers, which use a cryptographic algorithm and not with codes which are based on large, arbitrary code books. Figure 8.1 shows the nature of a cipher which employs two algorithms E and D to encipher and decipher information respectively. Encipherment operates on the plaintext using a cryptographic key (k) as parameter. Decipherment

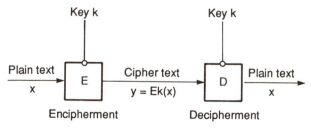

**Fig. 8.1** Notation for encipherment and decipherment.

uses the same key to operate on the ciphertext y and restore the plaintext x. The symbols x and y can either represent single blocks of data on which the transformation takes place or streams of data, for example, a stream of ASCII characters passing over a communication link. This distinction corresponds to two different types of cipher algorithm, a *block* cipher and a *stream* cipher. Later we shall give examples of both.

### 8.3.1  Attacks on a cipher

The cipher derives its strength, in part, from the key. If the secrecy depended on the algorithm alone, when this algorithm became known to an enemy, either by chance or by hard cryptographic work, it would be necessary to go back to the drawing board and design a new algorithm. All modern ciphers make use of a key so that, even when the algorithm is known, decipherment is not possible unless the particular key employed is known.

The possible number of different keys is important, like the number of 'differs' of the keys you use in a lock. If this is too small, a processor can be set to try all the keys and see which one generates a plausible plaintext. The total number of different key values is known as the *size of the key space*.

If nothing were known about the plaintext and it could therefore be any random set of digits, the most elaborate and fastest machine for searching through the keys would still tell us nothing. We must either have some knowledge of redundancy in the plaintext or, better still, some examples of actual plaintext with their ciphertext equivalents. In many cases, probable words or phrases from the plaintext can be guessed when their place in the text is unknown. With any of these clues about the plaintext an attack can be mounted, by searching through all the keys, in principle. A very large keyspace prevents the breaking of a cipher by simple key searching but it does not ensure that other, cleverer methods will not succeed. Only long experience with the practical breaking of ciphers can enable the strength of a cipher to be evaluated. Ultimately, there can be no guarantee because the whole range of possible methods of attack can never be enumerated.

### 8.3.2  The Data Encryption Standard

The US Government employs 'Federal Information Processing Standards' for its own use. In 1973, the US National Bureau of Standards announced that it was contemplating a standard for data encryption and asked for proposals. The response was disappointing but after a second call in 1974 a proposal from IBM was seen to have promise and, after some changes, it was published in 1975 as a draft. Eventually this was adopted in 1976 as the *Data Encryption Standard* defined by FIPS Publication 46.

The Data Encryption Standard or DES had a much wider influence than was originally intended. The US standards body ANSI made it a US standard and it was widely adopted, particularly in banking and financial services. Because of restrictions placed on the export of chips for the DES algorithms from the USA, its use outside the USA has largely been confined to banking and financial services. It became a *de facto* standard within this community.

The DES is a block cipher with plaintext and ciphertext blocks of 64 bits and it employs a key of 64 bits, but has 8 parity bits so that its keyspace depends only on 56 bits.

The structure of the algorithm is described in detail in reference [1]. It uses a combination of bit permutations and substitutions. By *substitution* we mean that an input field is used as an address to look up a table and produce an output field. The DES employs eight substitutions, each with a 6 bit input and 4 bit output. These are known as the 'S boxes'.

From the outset, the strength of the DES has been controversial. The size of its keyspace was criticised on the basis that a complete search through its keyspace could be carried out in less than one day by a machine containing one million devices, each able to test one million keys per second. The cost of such a machine was argued about but believed at the time to be in the region of $10M. The structure of the S boxes was clearly not random but the criteria for choosing these tables has never been published and the work done to develop the DES is classified. Many felt that this was unsatisfactory for a standard.

The whole idea of a published algorithm was a new one. Though the secrecy of the algorithm is not assumed to be essential to the security of a cipher, no cipher had been published, studied and discussed so widely and in such detail.

Studies in many places revealed interesting properties of the DES, including the existence of four *weak keys* and twelve *semi-weak keys* which give the cipher special properties. Interesting facets of the design of the DES continue to be discovered. However, none of these has shown any significant weakness of the algorithm beyond that which is implied by the size of its keyspace.

As information technology improved, the cost of a key searching machine decreased until, by about 1985 it became obvious that a searching machine would be within the bounds of expenditure by a large corporation or a large criminal organisation, assuming that the searching time was allowed to extend to about one month. This is a real threat when the same key is used for longer than this time. Multiple encryption using the DES with at least two different keys has been adopted by the financial community to overcome this problem. This is particularly relevant in key management for a large system where a master key remains in place for some time.

By 1988 the US had decided not to renew their endorsement of the DES for federal purposes. When this was announced in 1986 it caused consternation among the financial users but the US Government stated that it would not discourage use of the DES for financial transaction purposes.

At the present time, adoption of the DES or any other cipher, as an international standard has been stopped by a resolution by the ISO Council so there is no role for the international standards body in the development of a replacement. This leaves the financial community in an unsatisfactory position, particularly where the use of the DES for safe-guarding large payments is concerned. For small payments, with a good key management system, the DES is usually considered adequate.

The algorithm was designed for implementation solely in hardware on a special chip, according to the US Federal standard. The ANSI standard removed this condition and many systems now employ the DES im-plemented in software on a microprocessor. Because of its origins and the original intention to use hardware, the DES is not a convenient algorithm for software implementation and was outside the ability of the early smart cards. Recent developments in smart cards have made the DES more feasible so that it is now possible to implement the DES in software within the same smart cards. Where this is the algorithm required by financial transactions, incorporation of the cipher in a smart card is a useful contribution to security, enabling the card to store its own secret key and use it internally.

A replacement for the DES will probably be a block cipher with the same size of plaintext and ciphertext blocks in order to ease the change from DES to the new cipher. The key size would have to be larger, at least 64 bits. A replacement for DES should be designed to be easy to implement in hardware and software.

### 8.3.3.  Methods of using a block cipher

There is a limited number of applications for a cipher which can only handle 64 bits. A longer message could be broken into blocks of 64 bits, for example blocks of eight ASCII characters, and each block in turn could be enciphered. This simple method of enciphering a longer stream is not recommended for two reasons. Firstly, an enemy could extract pieces of the message made of a number of blocks and reuse them or rearrange them in messages of his own construction. Secondly, though eight characters may have a large number of potential values, in some contexts there

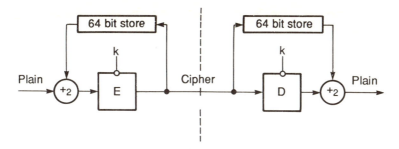

**Fig. 8.2** Cipher block chaining.

would be blocks which repeated quite frequently and could be recognised in ciphertext, so that methods used for breaking codebooks could be applied. For example, in English text the words 'of the', with the three spaces, when it occurs on the eight character boundary will probably be the most common block and can thus be recognised. In computer output, sequences such as all zeros or all spaces will easily be detected.

To overcome these problems a method of using the DES was introduced known as cipher block chaining (CBC) and this can be applied to any block cipher. It is illustrated in Figure 8.2 which shows the ciphertext produced by the sender added, modulo 2, to the next plaintext block. This chains each block to its neighbour, preventing the codebook analysis method and the separation and reassembly of blocks of text. When applying this method, the block which is modulo 2 added to the first plaintext block is called the *initialising variable* or IV. In many systems, a new IV is sent when each new key is distributed and the IV is kept as secret as the key.

Decipherment presents no problems, since the ciphertext can be stored and modulo 2 added to the plaintext of the next block. There is no synchronising problem in the sense that, whatever happens to modify or momentarily interrupt the ciphertext, provided the boundaries of the 64 bit block can be recognised at the receiver, the system will recover. For example, a single error in ciphertext results in completely random output for the plaintext block into which it is transformed. The same error goes into the 64 bit store and emerges in the following block, so two blocks of text are affected. After this, the error does not propagate any more. Even this amount of error extension can be troublesome. It can interact badly with error correction schemes and it greatly increases the average error rate on the line.

The CBC mode of operation is most useful where the information already has some imposed block structure, as in formatted messages. When these messages have been assembled in a store, the CBC encipherment can be applied to the whole stored message and then the result transmitted or stored, as required.

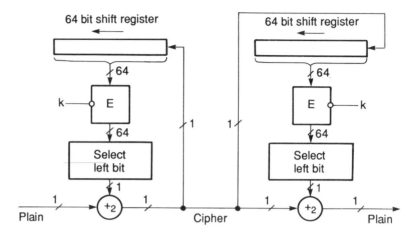

**Fig. 8.3** One-bit cipher feedback.

This mode of operation is not convenient when the natural unit of communication is smaller than a block. For example, much communication takes place in the form of 8 bit units or less, for example 8 bit ASCII characters or 5 bit telegraph codes. For these, the use of a 64 bit block cipher would mean that characters were held up, waiting for transmission, until the block was filled. Interactive communication between two operators would require that, after a short pause, the residual characters were padded out and transmitted. To satisfy these requirements, another method of using a block cipher was introduced which is *cipher feedback* (CFB). This is illustrated in Figure 8.3. At the top of the figure is a 64 bit shift register into which the ciphertext is introduced, bit by bit, from the right hand end. Thus it contains, at each end of the communication line, a record of the last 64 bits of the ciphertext. At each end of the line, this block is enciphered and from the result only the left hand bit is extracted. This bit stream is added into the plaintext to produce the ciphertext. Effectively, it is a random bit stream generated by feeding back the ciphertext itself. Note that the cipher algorithm is in this case used only in the encryption mode; in fact a reversible algorithm is not really needed.

Because this works on each bit as it arrives, it is completely transparent to the procedure used by the communicating parties, whether they are people or machines. It is normally used within the OSI architecture at the physical layer because it treats only the bit stream, without reference to its structure, and passes on each bit as it arrives.

The entire DES algorithm calculation must be made again for each bit, so if there is a limited speed of processing, this method will be slow.

As with the CBC mode, the shift register must be loaded before the process begins. In this case the initialising variable (IV) can be sent in clear over the line in a preamble. It can be in clear because the block is

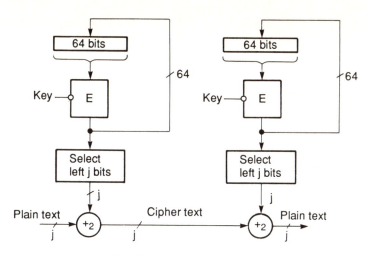

**Fig. 8.4** Output feedback.

enciphered before it is ever used. Another possibility is to send no IV but to transmit random data for the first 64 bits, which will ensure that the two registers are synchronised.

For this method, there is no synchronising because there is no 64 bit or other boundary to be observed. If an error occurs on the line or, for example, a bit is missed or an extra bit inserted, this causes an immediate error in the delivered plaintext and also affects the shift register during the next 64 bits. Consequently the error extension property of this algorithm produces 65 bits of error for a single error on the line. Error extension is a penalty paid for the advantage of any self-synchronising cipher method.

When the error extension property of the CBC or CFC is unacceptable, another mode called *output block feedback* or OFB should be used. This also enciphers by means of adding, modulo 2, a random bit stream to the plaintext and adding the same bit stream to restore the ciphertext. In this case, the bit stream is generated *independently* of the message content and is therefore unaffected by errors on the line. The method is illustrated in Figure 8.4. Here, the contents of the register at the top of the diagram are enciphered and put back in the register for each cycle of operation, at both ends. Provided the key employed is the same and the registers start with the same values, they should remain in step and generate, for each operation, the same 64 bit pseudo-random output. This can be added into the message stream as any size of unit that is convenient, for example as 64 bit blocks or as 8 bit blocks to match the character size or as 1 bit at a time. For the highest speed of operation, the whole 64 bit block is used.

With this mode of operation there is no error extension but synchronisation is a real problem. Each end of the line has a pseudo random number generator based on repeated DES encipherment of the register. The cycling of these two generators must be controlled by the clock rate of the line if they are to remain in step. A noise burst or short break in the transmission on the line can cause them to get out of step, after which the output at the receiving end is a random stream and communication is lost. In a practical system, there must be a means to *detect* loss of synchronism and *restore* it. For detecting loss of synchronism, there must be some redundancy in the plaintext which can be measured easily by the receiver. There must also be a return channel to tell the sender that synchronism has been lost, a protocol for exchanging a new starting value for the register and a method of restarting the two ends simultaneously. The starting value can be transmitted in clear across the line because it is enciphered before it is used. Some systems of this kind interleave information with the enciphered data which is accumulated to form a new starting value. It is possible to resynchronise at intervals, whether synchronism is lost or not and in this way be sure to restore synchronism, at least after an interval. Some systems resynchronise whenever a certain pattern is recognised in the ciphertext. However this can cause troubles when errors on the line make one end believe the pattern has occurred and the other does not. Maintaining synchronism in these stream ciphers is a difficult problem, particularly if they must be transparent to any pattern in the plaintext and therefore not dependent on plaintext redundancy to recognise loss of synchronism.

The three modes of operation that we have described, CBC, CFB and OFB, between them cover most of the applications of a block cipher for data confidentiality. We shall return later to its application for data integrity. These modes of operation are applicable to any block cipher and therefore form a rather general and useful aspect of cryptographic technique.

Cipher feedback can be applied with the feedback path carrying more than one bit on each cycle of the device. Apart from the one bit cipher feedback we have described, the next most common form is 8 bit cipher feedback in which 8 bits are selected from the left hand end of the cipher output for addition into the message stream. At each operation, the 8 bits of cipher text produced are shifted into the shift register, so that after 8 operations the register contents have been completely renewed. This is a convenient way to encipher an asynchronous transmission with 8 bit units. Other feedback widths are possible but are rarely used.

### 8.3.4 Key management

Under the heading of key management we include the entire life cycle of a key from its creation, through its distribution and use to its eventual

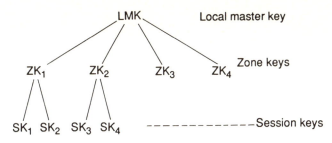

**Fig. 8.5** A key hierarchy.

destruction. Keys should be chosen randomly within their total range (key space) and the ideal is to employ a true random number source such as a circuit which amplifies and samples random noise. A second best alternative is to generate a pseudo-random sequence using a cryptographic algorithm and some starting values called 'seeds'. Where random numbers must be generated within a smart card this is normally the method used.

For the distribution of keys, storage in a smart card for carrying to the place where they are to be used is effective, assuming that the smart cards are physically able to protect them from reading.

The principle employed in many key management schemes is that the same key should not be used too many times or over a very long period. For this reason, the keys that are used for enciphering or providing integrity checks on data are *session keys* which are changed frequently. These session keys can be transmitted over a communication line from a convenient central source provided that a master key is available to encipher them in transit. This is the first stage of a *key hierarchy* in which one key enciphers another. In practice, hierarchies of several layers are employed as illustrated in the example of Figure 8.5. This shows a *local master key* (LMK) at the top of the hierarchy which is used to encipher a number of *zone keys*. Each zone key is employed between a pair of units (or perhaps more) which have direct links with one another. The zone keys have a long life so they are used to encipher the session keys which actually carry out the encipherment of data or the integrity checking.

The top level key or LMK is used locally to encipher all the zone keys which are employed at that location. The purpose of this encipherment is to allow the zone keys to be stored outside the protected environment. This is necessary where there is a complex key hierarchy and the storage capacity of the cryptographic module is limited. These days, storage limitations are less frequent but external storage allows more than one security module to share zone keys if they all have the LMK stored in them. Zone keys are the master keys which operate between distant units and must be exchanged between these units. These can be carried in a smart card or similar protected unit. The encipherment of sessions keys using zone keys is part of the automatic key distribution carried out over the communication lines.

The full details of a practical key management scheme are complex and should use sequence numbers and identifying information for all the session keys that take part. Several methods of linking these data together have been employed; and the terms *notarisation* and *offset* will be found in the literature and the standards on key management. Among the many aspects of the design of secure systems, key management is the one that is more often specialised to particular applications. In recent years, smart cards have taken an increasing part in these customised key management schemes.

There is one key management scheme published as ANSI standard X 9.17 and (modified) as an international standard ISO 8732. It is intended for the management of keys in wholesale financial transaction systems. Taking into account all its details it may not be suitable for use outside this range of application but the terminology of its messages and the schemes of encipherment it uses are found in key management schemes for other applications. Because of its complexity it is best referred to in the original standards documents.

The key hierarchy is a useful concept because it enables the working keys to change while reducing the frequency at which the top level keys must be transported. It can never eliminate the need for physically transporting some keys before a system can be started up. Because these top level keys control all other cryptographic keys, their security against illicit reading is vital and there are risks in physically transporting any secret data. A recent development in cryptography is the *public key cipher* and this can make a radical change in the top level key management.

### 8.3.5  Public key cryptography

The principle of cryptography shown in Figure 8.1 employs a secret key k as a parameter in both the encipherment and decipherment algorithms. This is the way in which individual pairs of users can ensure that their communications are produced from others who do not know the secret key. The principle of a secret key has been accepted as fundamental by cryptographers for more than a century and the majority of today's ciphers are of this kind. We call it a *symmetric* cipher because the knowledge of the sender and receiver is the same and because, with this secret key, encryption can be applied to information passing in either direction.

A remarkable new idea was proposed by Whitfield Diffie and Martin Hellman in 1976, who noted that only the receiver of the information, who needs to decipher it, requires a secret key. They proposed a type of cryptography in which the key used for *encipherment* was not secret. Clearly, if there are two different keys they must be closely related in order that encipherment and decipherment shall be relatively inverse. If the encipherment key is not secret it is essential that knowledge of this key does not betray the secret decipherment key, so the relationship between the two keys is an unusual one. The tricky nature of the functions used in public key cryptography is even clearer if we consider that the whole process of encipherment, its algorithm and its key, can be made public without betraying the processing of decipherment. Anyone can perform the translation from plaintext to ciphertext but it requires secret information to perform the inverse function.

Functions of a kind which are easy to perform in one direction but for which the inverse is extremely hard to calculate have been known for some time under the name *one way functions*. Encipherment is a special type of one way function because it can be inverted if a secret key is known. This is sometimes called a 'one way function with a trap door'. After the first description of this principle, it was two years before a practical scheme was published by Rivest, Shamir and Adleman and this crypto system, the *RSA* cipher, is still the most practical public key system. Figure 8.6 shows the principle of a public key cipher.

As a starting point for generating the two keys we assume a seed value *ks* which has been chosen at random. From this value, using published algorithms F and G, the two keys are generated. *ke* is the encipherment key and *kd* the key used for decipherment. The key *ke* is described as a public key though there is no necessity to make it public. If an enemy does learn its value, this still does not reveal the secret key *kd* which must be retained and protected by the receiver of the information. The figure shows that the receiver should also control and protect the device which finds *ks*, generates *ke* and *kd* and uses the latter for decipherment. This is an *asymmetric* cipher and if two-way communication is needed the other party will have to generate his pair of keys and send out the public key for the encipherment of information on its way to him.

If the public key is really made public it allows anyone to encipher data and send it to the owner of the corresponding secret key. For a very large population of users of which anyone may need to send information to any other, this is a valuable feature because secure communication requires only that all parties shall know the public keys and then secret communication can take place between any pair. For $n$ such users only $n$ key pairs have to be generated and their public values placed on a list which is made known to all. With asymmetric or secret key cryptography, the number of keys needed for this situation would be $\frac{1}{2}n (n - 1)$.

Someone who receives a message enciphered with his public key can decipher it and, by its redundancy, know that it was enciphered with the correct key. This tells him that the sender knew his public key but if this

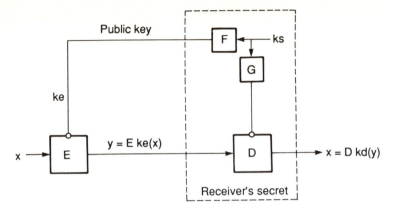

**Fig. 8.6 Principle of the public key cipher.**

key is widely known it does not identify the sender. In other words, there is no authentication of the received message. By contrast a symmetric scheme in which each communicating pair shares a secret key authenticates the sender of the message. There are many ways to provide this *source authentication* using public key cryptography. Suppose, for example, that each time a message is sent a random number field is included in the message. The reply from the receiver of the message also quotes this random number (as well as giving a new random number so as to continue the process). After the first message has passed, this procedure provides authentication of the origin of each message. The sender of each message knows the public key of the intended receiver and only that receiver can decipher the message. Therefore if his random number comes back in reply, it must have been deciphered at the destination, so the source of that reply must be the owner of the corresponding secret key. This authenticates the source of the reply. This simple procedure can maintain source authentication with public key cryptography. There are alternative ways to provide authentication using *digital signature*, described in 8.4.

The distribution of secret keys for symmetric ciphers is a great problem. With public key cryptography, the numbers to be exchanged do not require secrecy; they do require to be authentic. If an intruder could plant his own public key in place of the true one, he could intercept the messages and decipher them. But the intended receiver would find them undecipherable so the intruder has achieved little unless he can re-encipher the messages with the true public key and pass them on. This kind of active attack, which must intercept every message and re-encipher it without detection, is extraordinarily difficult and for many purposes could be regarded as impossible, nevertheless authenticity of the public key is an important security factor.

In a large community with a common set of public keys, this can be handled by installing a public key registry. The participants' public keys must be correctly registered and the users' access to the registry must be protected against falsification. Good solutions have been found to these problems by using the digital signature principle and by developing a good protocol for the key registration.

Public key operations involve heavy processing, in particular the operation of decipherment. It seems a fact of life that public key cryptography entails a heavy load of processing. For the first decade of this new technology the processing load imposed severe limitations on what could be done with it. Typical microprocessors took a minute or more for the decipherment operation and this could not be used directly for practical cryptography. Much faster special chips could have been made but none became available. The first application was therefore to ease the problem of distributing secret master keys for a symmetric key hierarchy. Compared with the physical movement of a secret key from one place to another, a minute or more of processing was convenient and much safer. The advance of technology has changed this so that now cheap microprocessors can perform the decipherment function in about a second. This still does not produce a good data encipherment rate but there are many transaction systems where it is adequate. Furthermore, the nature of the RSA cipher enables encipherment to be performed about 50 times as fast as decipherment by making a suitable choice of keys.

The first chip design for RSA operations came on the market in 1987 and gave an order of magnitude speed improvement over the fast microprocessors. This enabled bulk handling of RSA decipherment and made the use of public key cryptography without a symmetric cipher practical for the first time. The incorporation of RSA functions into a smart card is the next step. This can make a qualitative change in security and is the subject of Chapter 6 of this book.

Chip designs for much faster RSA processing have been proposed but commercial activity has been slow. The investment made in the development of DES chips was not very profitable for most suppliers. This has discouraged the design and development effort needed for chips for RSA operation. The gradual development of technology will overcome these limitations in a few years.


## 8.4  DATA INTEGRITY

By verifying data integrity we ensure that no unauthorised change has been made to information, that is to say we verify that no active attack has taken place. At one point the information is cryptographically 'sealed' in such a way that any later change can be detected. Two processes are involved, the first which creates extra information at the time that the information is known to be correct and the second which allows the integrity of that information to be checked at a later stage.

Data integrity checking can be applied to communication by using the first process before sending and the second process when information is received. Alternatively it can be applied to stored data so that the reader of the data can verify its integrity, meaning that it has not been changed by an unauthorised person while it was in store.

### 8.4.1  Principle of integrity checking

The principle of the cryptographic method is shown in Figure 8.7 in the context of data communication. A message M is transmitted from source to destination. Together with the message is sent a number A which is called an *authenticator*. More strictly, it should be called a *data integrity check field*. The value of A is derived from the whole of the message using a calculation which employs a secret key k as parameter. At the receiving end, the same calculation is performed on the received data using the same secret key and the result is compared with the incoming value of A. If these are equal, the integrity of the message M has been verified. If an error appears at this point, it may be due to deliberate unauthorised changes on the line, or it could be due to an error or using the wrong key.

The algorithm used is cryptographic, in the sense that if a number of examples of messages together with their authenticator values are known, it must not be possible to deduce the value of $k$. The value of A must also depend on every bit of the message so that no part of the message can be altered without detection. In general, authenticators have been developed from cryptographic algorithms but this is not essential because, unlike a cipher algorithm, no corresponding inverse is needed.

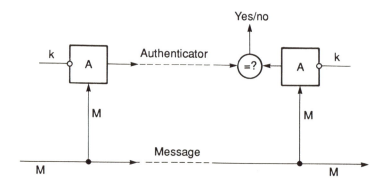

**Fig. 8.7** Principle of message authentication.

The size of the authenticator field is not dependent on the message size so the authenticator, although it depends on the whole message, is usually of a fixed size, such as 32 bits. The main criterion is that the equality on which the verification of identity depends should be very unlikely to happen by accident. In fact, a 16 bit authenticator with a probability of accidental agreement of 1/65536 would be sufficient for most purposes, but 32 bit values have become well established.

### 8.4.2 Prevention of misuse and replay of messages

If the same key is used for some time, entire messages might recur, and then their authenticator values would be identical. Making use of this, an intruder could repeat an old message and have it accepted as genuine. This would be serious if the message is a payment. Another possibility is that a message intended for one purpose could be misused for another and the authenticator would still match; although data integrity in a narrow sense has been preserved, this would not be a safe system.

Both these problems can be solved by careful choice of the format of the message which is authenticated. To avoid misuse of the message, it should contain full identifying information such as its source, persons authorising it and the identity of any other persons or accounts involved. Provided this information is comprehensive, misuse of the message becomes impossible. For example, all key management messages should contain the identity of the source, the purpose for which the key is to be used and any other information which limits its range of use.

To prevent the reuse of a message for its original purpose, all messages should carry a sequence number of some form. If the messages are passing between two entities only, a sequence number can be maintained for this one source and destination. The receiver checks the sequence number. If one source sends messages to a number of destinations, the destinations can check the increasing sequence but not that the sequence is complete. This at least ensures that no message can be reused but does not guard against the deletion of a message in the sequence.

Where deletion of messages is not a significant risk, the sequence number can be replaced by a time and date stamp. This is very convenient when a large population can be either senders or receivers of messages, as in electronic mail.

### 8.4.3 Algorithms for data integrity checking

Data integrity checks for financial messages are, in most cases, based on the DES algorithm. A method of using this algorithm to calculate an authenticator field is shown in Figure 8.8. This corresponds with the

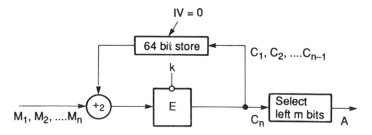

**Fig. 8.8** Authentication using the cipher block chaining mode.

mode of operation we described in section 8.3.3 as cipher block chaining. When CBC mode is used for calculating an authenticator, the IV is zero. The message is divided into sections of 64 bits, padding out the final block to 64 bits if necessary. (Padding out with zeros is acceptable.) The first block is enciphered using the secret key, then added modulo 2 to the second block and enciphered again. This process is continued until all the blocks have been used. From this procedure, a 64 bit number is produced and, by convention, we use the left hand 32 bits of this result.

The usual name for this type of authenticator is a *message authentication code* (MAC). The terminology is so well established that the term MAC normally denotes an authenticator calculated in this way with the DES algorithm. The current ISO terminology reserves the term *authentication* for the authentication of peer entities, notably user authentication. It does not introduce a satisfactory word for the concept of *data integrity check function*. Until a useful single word is adopted it seems acceptable to use the word *authenticator* for this quantity. The term MAC is less satisfactory because the 32 bit field is not strictly a 'code'. It has been so widely adopted that use of the acronym MAC is unavoidable.

There is no published integrity algorithm that has been used as widely as the MAC. When it was considered as an ISO standard for financial messages, the inconvenience of a software implementation of DES was recognised and an alternative algorithm developed specifically for use with mainframe computers was adopted as an alternative. This so called *message authenticator algorithm* (MAA) employs a 32 bit integer multiplication and is therefore not only suitable for mainframes but also for microcomputers of more recent design. In this algorithm, messages are divided into blocks of 32 bits with padding of the final block if necessary. A 64 bit key is used from which an initial calculation produces six 32 bit numbers. Two of these numbers form initial values for the calculation, two of them are used during the cycles which successively employ the 32 bit blocks of the message and two of them take part in a final two cycles of the calculation called the *coda*. The running calculation operates on and produces two 32 bit numbers and, at the final step, these are added modulo 2 to form the authenticator value. The full definition of the MAA algorithm is contained in ISO 8731 part 2. If the message is longer than 1024 bytes (256 blocks of 32 bits) an authenticator is calculated from the first 1024 bytes then prefixed to the next 1024 bytes and the calculation of the authenticator is started again. The authenticator (derived from the final block) is the authenticator for the whole message.

There are many other data integrity algorithms in use but some are proprietary algorithms which have not been made public. Among these is the algorithm used by SWIFT. In smart cards, a proprietary algorithm known as *Telepass* has been used. As the capability of the microprocessor in smart cards improves, it may become possible to adopt the ISO algorithm MAA.

### 8.4.4 Digital signature

The DES based algorithm and MAA employ a secret key which must be known to the entity which calculates the authenticator value and also to the entity which checks it. If this key is to be kept secret it must not be made widely known; therefore this type of integrity check is primarily used bilaterally. It needs the prior distribution of a secret key. The technique protects the information against falsification by any third party who does not know the secret key. It does not protect either of the two entities against dishonesty of the other. The receiver of the information could alter it and since, with a knowledge of the secret key, he could generate a valid authenticator value, this does not protect the information against changes. This technique would be worthless as a method for protecting the information in a bank cheque, because the receiver has a motive for falsifying it.

Because the receiver could alter or forge a document the sender can falsely accuse him of doing so. After having sent a message, such as payment, the sender may wish to deny having done so and the evidence of an authenticator value calculated with a secret key gives no protection for the receiver against this charge. If this kind of dispute between the sender and receiver of a message is to be resolved, they must not share exactly the same information, in other words the *asymmetric* nature of public key cryptography is required in this situation.

A simple change in the operation of a public key cipher can provide a type of authentication which meets the requirement exactly. This is illustrated in Figure 8.9 where the algorithms and notation are those of Figure 8.6 and the operations have been changed around so that the message x is first subject to 'decipherment' then to 'encipherment'. For the RSA algorithm the two operations D and E are, in fact, identical and their different effect is due only to the keys they use.

*Integrated Circuit Cards*

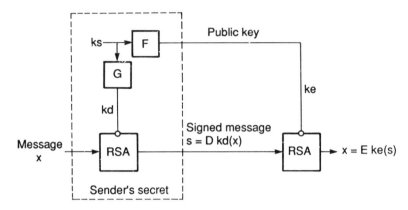

**Fig. 8.9** Principle of public key signature.

The message produced in this way is transformed but is not secret because anyone, with the aid of the public key, can restore it to its plaintext form. The existence of the restored message (recognisable as such by its redundancy) is proof that it came from the sender who owns the public key. Therefore, if the public key is authentically known, the transformed message is evidence of the message's origin. In effect, the value *s* and the public key *ke* could be taken to a third party for arbitration. The existence of the restored message text *x* as a function of these quantities, together with the known identity of the owner of the public key, is proof of the origin of the message. Figure 8.9 shows the principle of the method but in this form it would be inconvenient. The receiver of the message would need to store the 'signed message' *s* but for all practical purposes only the plaintext x is useful. For very long messages this would be particularly inconvenient. Also, it does not correspond with the way in which we understand signatures, where the signature is a relatively short piece of information added to the message. A more convenient method is shown in Figure 8.10. Here, the message is sent in plain; a signature consisting of one RSA block is formed by applying the signature process to the quantity H(M) which is a one-way function of the whole message. At the receiving end, the same one-way function is formed and is compared with the quantity generated by applying the RSA process to the received signature.

There are many ways to calculate the one way function H, some using the DES algorithm and others using RSA operations. No standards have yet been established in this area. With the signature scheme of Figure 8.10, messages of indefinite length can be signed with one RSA block, which is typically of length 64 bytes.

Encipherment with a public key cipher does not, by itself, provide authentication of the message. The signature process, in either of the

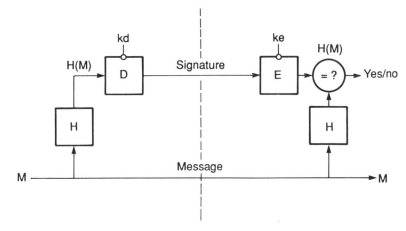

**Fig. 8.10** Signature separated from the message.

forms we have described, does not provide message confidentiality. By comparison, an asymmetric cipher provides encipherment and, with certain precautions in the format of the message, it also provides a measure of authentication. Public key cryptography separates these two properties into two different processes. If both encipherment and authentication are needed, the secret keys of the sender and the receiver are both involved. The message shown in Figure 8.10 can be enciphered and the encipherment can include the signature block or not, as required. Alternatively, the simple protocol described in section 8.3.5 can provide authentication for an exchange of messages.

Both digital signature and public key cryptography depend on the authenticity of public keys. In a large group who use it for message interchange all the public keys are held in a central registry. For distribution of these keys the messages can be signed using the key of the registry so that each user need only have an authentic value for the registry public key to make the whole key distribution scheme secure.

One of the main virtues of digital signature as an authentication method is that the public key can be made known as widely as required. Therefore messages authenticated in this way can be checked by any party which requires to verify their authenticity. This is important when a message passes through a number of hands in the course of its processing or when a single message is to be broadcast widely.

A practical use of digital signature was held back by the computational complexity of public key cryptography but the introduction of fast processors now makes it practical for nearly all purposes, soon including its use in smart cards. By using a short exponent in the RSA scheme the process of checking a digital signature can be made very fast so that only the generation of the signature requires the full length RSA operation, this taking about one second in a fast microprocessor.

## 8.5   USER AUTHENTICATION

In a communication system with a layered architecture (such as OSI) two entities which communicate with each other are in the same layer and are called 'peer entities'. Authentication is the means by which two communicating peer entities each establish each other's identity in a secure manner. This is the general case of *peer entity authentication*.

As an example, consider two cryptographic modules which are used to verify the integrity of data passing between them. They do this by means of a secret key which they share in common. If information is checked with the secret key and proves to be genuine, it must have arisen from a place where a secret key is known so that the appropriate check function (authenticator) could have been produced. Therefore, if the key management is safe, this also provides peer entity authentication. The receiver knows not only that the information received is genuine but also that it arose from the corresponding module (peer entity) since it is the only one holding that secret key. Thus peer entity authentication in this case is a natural consequence of secure key management. Indeed, data integrity is of very little value if it is not combined with authentication of the source of the data. Data which has not been falsified during the communication process is of no value if it was falsified by a masquerading peer entity.

The type of peer entity authentication which is of most importance in practice is authentication between a *user* and a computer system with which the user is interacting through a communication network. If the users are mobile and can appear at different terminals or at different ports on the communication network, authentication of the terminal is not enough and there must be authentication of the user himself. Generally this must be 2-way authentication so that the user is not fooled into interacting with a masquerading central system which could cheat him into believing that a transaction had been made or obtain information from him by this trick.

### 8.5.1   Existing methods

By comparison with the security which can now be obtained, user authentication is often no more than the presentation of a password, which can be very insecure. The best password systems can have a reasonable level of security. These use passwords which change at every access attempt and these passwords are presented both by the user to the system and by the system to the user.

Passwords came into use with the first interactive systems because they could be implemented with nothing more than a simple program at the host. Far too much trust was placed in this insecure method with the results that we now see in the activities of 'hackers'.

In an attempt to provide better user authentication or to supplement the security of passwords, a number of schemes for 'biometric' user authentication have been developed. These attempt to measure a characteristic of the user which he cannot easily change such as a fingerprint, shape of hand or blood vessels on the retina. Alternatively they use an action by the user over which he has little conscious control such as the dynamics of his manual signature or his voice. Biometric methods have been improved steadily but they all have perceptible error rates either by rejecting genuine users or accepting false ones. There is a trade-off between these two types of error but some level of error seems unavoidable.

### 8.5.2 User authentication employing a smart card

A cryptographic exchange of data can provide effective peer entity authentication but if one of the entities is a person, he must hold a cryptographic device on which the processing can take place and which he will ensure is not lost or stolen. This is an example of a token (something which a person holds) being used to authenticate the user. Tokens such as magnetic striped cards, which have no processing power, can easily be read, copied or forged and are insecure. To be effective, a smart card is needed, which can protect its key physically and also contains the cryptographic function for performing the authentication protocol.

There are several protocols that can be used. The most commonly used protocol is shown in Figure 8.11. The host sends a random number R (the challenge) which is processed by the card using the secret key k and returned (S) to the host. The host also has the secret key and can perform the same calculation, comparing its result with the one received from the smart card. If these agree the smart card is taken to be authenticated and, by implication, the user also. A card can be protected against unauthorised use by someone who finds or steals it by requiring a password or personal identification number to be given before it will perform its cryptographic exchange. The password can be entered either by a keyboard on the token itself or from a separate keyboard through its electrical interface. The latter raises the problem of a false terminal which can capture the password, though this is of little use without having the token as well.

The challenge and response procedure can use a cipher algorithm in which the secret key held in the card is used to encipher the challenge and produce the response. The inversion property of a cipher algorithm is not needed so any cryptographic function, such as a one-way function could be substituted.

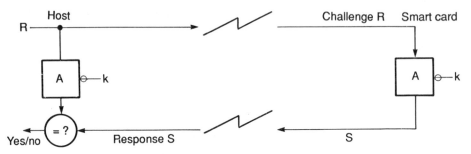

**Fig. 8.11** Challenge/response method of authentication.

The token need not have an electrical interface with the terminal if it has a keyboard and display. The user can enter the challenge into the token and read the response from it. There are on the market a number of devices resembling hand-held calculators which work in this way. It requires extra effort from the user, who must read the challenge from the screen of a terminal and enter it into the token, then read the response from the token and enter it into the keyboard of the terminal. The advantage of a token without an electrical interface is that it can be applied to the hardware of the terminal, in fact it can be used by telephone where the numbers are spoken from one person to another.

In order to reduce the effort needed with the non-interface token the transfer of the challenge number can be dispensed with if the response is made to depend on a changing number stored in the token for which a corresponding number is maintained at the host. Figure 8.12 shows one such principle. Registers at the host or in the card hold a value $x$. This is processed using the secret key $k$ to generate the response S. Then this value becomes the $x$ value for use on the next occasion. This simplifies the user's procedure but ties the token to just one host which is, of course, the most common requirement. Other devices on the market employ a quartz crystal clock in the token to keep it in synchronism with

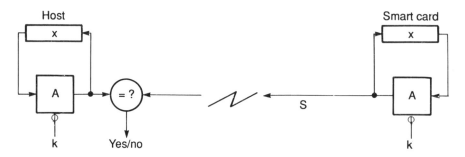

**Fig. 8.12** Authentication with a stored value.

the host, enabling one token to be used with a number of hosts. The clock's time and date form the 'challenge' from which the cryptographic response is sent via the user, who transcribes it to the terminal, to the host. Because the clocks will slowly lose synchronism, each active host must make small corrections and be prepared to try a few challenge-response pairs to determine whether the token is genuine.

Hand-held devices for cryptographic authentication of users are valuable chiefly because they can be fitted easily to existing systems. Where a new system is being designed, the electrically coupled smart card is preferable.

### 8.5.3 Authentication by digital signature

Cryptographic exchanges using secret keys can provide excellent security but a single token's use is necessarily confined to the corresponding hosts which possess these secret values. A much more general method of user authentication can be developed using digital signature. This is the basis of the signature token described in Chapter 6.

The principle is that the token generates a public key pair and holds its secret key protected inside it. Any host wishing to use this token for authentication is given the public key. The cryptographic exchange consists, as before, of a challenge and response. The response consists of the 'digital signature' of the challenge and can easily be verified with the use of the public key. It is, however, important that the challenge cannot be used to create a message which could falsely be said to have been generated by the owner of the token. The protocol and an agreed message format prevent this.

Because the information used by the host consists only of a public key, the number of hosts able to collaborate with a given token is unlimited. In practice, the useful transactions which could follow this authentication stage must be agreed between the host and the token. Therefore it is likely that these tokens will be specialised to certain types of application, such as banking. Within this general area, provided there are agreements on protocols and message formats, the range of possible transactions could be large. For these transactions, a number of banks and financial institutions could collaborate with a single token held by the user. The digital signature principle overcomes the security problems of this arrangement.

### 8.5.4 The Fiat Shamir authentication principle

Development of a practical signature token is well advanced but there would be an advantage in a comparable authentication method that could be performed with less computation than the RSA calculations require. In particular, calculating the digital signature is computationally complex, while verifying the signature can be simplified by using a small public exponent. Fiat and Shamir have described such a scheme which is based on the properties of square roots in modular arithmetic.

Like the RSA method, it employs a modulus which is the product of two primes. For the public key, only the product is made known and the numbers are large enough that factorisation is practically impossible. Given such a composite modulus, the properties of square roots that are employed are as follows. About one quarter of all numbers have square roots and these are called *quadratic residues*, QRs. Nearly all the QRs have four square roots which are arranged as two pairs. If $m$ is the modulus a pair is $x$ and $m - x$, rather like positive and negative roots of ordinary arithmetic. It is convenient to make the factors of the modulus of the form $4n + 3$ since it makes it easy to discover if a number is a quadratic residue and, if it is, to find a square root.

The property used in this authentication method is that, if the factors are known, it is easy to calculate the square root of a QR but if the factors are not known, finding the square root is as difficult as factorising the modulus.

Fiat and Shamir developed a number of schemes but, for simplicity, we shall describe the simplest of these methods, though it is not the most efficient.

When the card is first established, it finds a suitable modulus m as if it were an RSA token. The user then generates a message containing the identifying information which the card will authenticate to any enquirer. If necessary, a one way function of this message can be used to reduce its bulk. The resultant number must be made into a QR, which can be done by having a small field the value of which can be adjusted to make it a QR. The card is then able to calculate a square root of this identifying information. That is, it finds a quantity u such that $u^2 = v$, modulo m, where v is the identifying information.

The problem is for the card to prove to an enquirer that it contains v without revealing that value. This is handled by a repeated dialogue of the following kind.

(1) The card finds a random value r and sends $x = r^2$ and $y = v/x$ to the terminal which checks that x.y = v.
(2) The terminal sends a one bit decision $e$ to the card to decide the next move.
(3) If e = 0 the card sends r to the terminal which checks that $r^2 = x$. If e = 1 the card sends s = u/r to the terminal which checks that $s^2 = y$.
(4) The terminal checks that $r^2 = x$ or $s^2 = y$ according to its choice of $e$.

If both r and s had been sent, the secret u would have been lost, since u = r.s. By offering to send either, but not both, at the choice of the terminal, the card shows its confidence that it knows both, hence knows u.

From the terminal's viewpoint, either x or y might have been constructed as $r^2$ where r is a chosen value so the outcome could have been a piece of luck with probability ½. For this reason, the above dialogue must be repeated with different values r many times, so that the probability of the card succeeding by luck is very small. For example, 20 iterations would be sufficient.

These 20 operations are still much less onerous than a single RSA exponential calculation since the multiplications and divisions involved are very small tasks compared with exponentials. The communications load for the 60 or so numbers must also be considered. This is unimportant when the authentication is a purely local function.

In an improved scheme of the same kind, the terminal makes five binary choices at each stage instead of just one and, in this case, four cycles of dialogue gives the same low probability of the card succeeding by luck. This also reduces the communications load.

There will probably be useful applications for the Fiat and Shamir authentication scheme but, unlike the RSA procedure, it cannot function as a signature method. A record or 'transcript' of the dialogue is of no value in proving the authenticity of the card after the event because it could be forged by deliberately checking always the easy option, since x and y cannot be distinguished. Fiat and Shamir have produced a true signature scheme using the same principle which offers a saving of computation compared with RSA signature by a factor of about 20. Unfortunately, the amount of data stored in the card and the size of the signature are both increased. Since the original publication, many new versions of this 'zero knowledge' protocol for authentication and corresponding signature schemes have appeared, so that various trade-offs between computational complexity and sizes of keys and communicated data are now available.

## 8.6 THE FUTURE OF CRYPTOGRAPHY IN THE SMART CARD

Improvements in semiconductor technology will allow closer packing of logic and storage devices on chips, higher processing speeds and lower power consumption. This will remove some of the constraints on complex cryptographic algorithms.

Current technology allows the use of the DES algorithm in a smart card of standard bank card dimensions, in spite of the rather poor match between the nature of the DES algorithm and the simple processors employed in smart cards.

The replacement of the DES algorithm by a more secure cipher and one which is better adapted to be implemented in software or firmware is an urgent requirement and is technically possible but is unlikely to be worked out in the public arena and made the subject of a national or international standard. Where the requirements of banking and other financial services are concerned it is possible that a suitable algorithm for limited use within this community could be developed but there is no organisation established internationally which can easily take up this challenge.

In the next few years, the implementation of the RSA cipher in compact *signature tokens* will be feasible. There is a real possibility for the use of the RSA cipher to overtake the DES and become a *de facto* international standard, regardless of the pressures on ISO. Digital signatures employed within user-held tokens, as described in Chapter 6, can provide a very effective mechanism for personal transactions of all kinds. Though the Fiat Shamir authentication scheme is attractive, the more general purpose nature of the RSA algorithm as cipher, signature and one way function, makes this in the longer term, more significant.

The security of a user token depends not only on its cryptography but also on its physical security and the way in which information enters and leaves it. In this respect, the thin card of the present banking standard may not remain the best choice. A keyboard on the card and a display can improve the security by verifying to the user the exact nature of the transaction being performed, providing the user with information authenticated in the card and receiving instructions and the password (PIN) directly instead of routing it through a less secure terminal. Also, the coupling of the token to the terminal or communications interface must eventually avoid electrical contacts. For all these reasons, a slightly thicker and rigid format for the user-held token seems likely to emerge.

The technology will evolve and, whatever standards eventually emerge, the interplay between cryptography and other aspects of security in the smart card of the future is an interesting prospect.

## REFERENCES

[1] Davies, D. W. and Price, W. L. (1984) *Security for Communication Networks* John Wiley.

Chapter 9

# Smart Cards –
# the User's View

## DR CATHERINE P. SMITH

### (DOICA Ltd)

*New technology must provide benefits for the end users.*


## 9.1 INTRODUCTION

The increasing miniaturisation of electronics and the developing resilience
of assembled units make the smart card potentially one of the most
exciting and interesting of the gadgets ever to arise from the technological
age. The implications for the consumer are fascinating. Imagine, within
thirty years, an object the size of a credit card, carrying the individual's
internationally recognisable unique passport identification, full medical
history, total financial record – and incidentally the ability to authorise
new payments and transfers – and a complete personal diary and time-
planner, all under the individual's secure personal control. Add to that,
that some versions even incorporate a telephone, and next year's model
will have a television screen on the reverse.

Fantasy? Well, try telling the average consumer in 1958 that within 30
years his record player would have quadrophonic sound, that 20 songs
would be on an indestructible disc a few inches across, and that he could
listen to it in total privacy in the tube as he commuted.

This book is not, of course, addressing what the marketplace is going to
be like in 30 years time, but rather how the average consumer will react
over the next five to ten years as the new applications of the smart card
become available. Indeed, many readers of this book are seriously concerned about how to make these applications not just available, but a
commercial success.

A lot of the experimentation with smart cards, and much of the current
interest, is in their use for financial services. The financial institutions are
keen to realise a return on their investment into smart card research.
Understanding the consumer is essential if they are to achieve this return.
This chapter provides a résumé of existing market research results, and
indicates how reaction might change over the near future.

In most bank research, smart cards have been used as payment cards, but more recently, banks have begun to use smart cards as access tokens to sensitive financial services, such as corporate cash management systems.

Most of the experimentation with smart cards as payment cards has taken place in France. Smaller experiments have taken place around the world, with the most significant being the Mastercard experiment. VISA is at present involved in extensive trials of supersmart cards. In each case, cards have been distributed across the whole range of traditional payment card users. In each case, consumer reaction has been equally positive (or equally indifferent) across the range of users.

In fact, these payment card experiments have demonstrated that the payment card user is virtually indifferent to the technological mechanism which makes the payment. However, he is very interested in the fact that he can use this particular paycard (and not others) in special locations − including taxis, telephones, and through the videotex system. But he responds to the extra uses − not to the technology.

Early experiments with smart cards in banking were focused on local geographical areas, or common interest groups. Bank experimentation with smart cards began in France, following a decision in 1979. Field trials started in 1982. The object of the early experiments was principally to try out the technical capabilities of the cards, though the trials clearly indicated that cardholders were interested in the services offered to them, and not in how they were offered. Two years after the French trials started, the Norwegian experiment began. Since the technology was already being tested in France, the project was able to focus on issues other than the purely technical ones. [1]

One of the interesting observations from the Norwegian experience was the interested response of consumers wishing to take part in the project. Out of 19,000 bank account holders who were invited to put themselves forward as card holders in the experiment, there were 3,700 positive responses in the first week! 5,000 cards were issued. The customer base very closely matched the general population of the area, giving the lie to the suggestion that it was young people who would be drawn to the experiment by their interest in the technology. A concurrent experiment was run with Telecards for payphones, with 5,000 smart cards being sold for use in public payphones. These phones could also accept the bank smart cards.

Within six months of the start of the Norwegian experiment, nearly 70% of cards had been used, with active cards used about three times a month. The highest proportion of those who did not use their card was amongst retired people. However, the most active customer was a lady in her 80s, who had used the card 130 times in the six month period. Such customer opinion as was gathered suggested that customers were happy with the card, and in particular were interested in getting additional services through the card.

A third experiment, in 1985, this time amongst a common interest group rather than within a strict geographical area, took place in Bormio in Italy during the Valtellina World Downhill Ski Championships. The experimenters in this case thought of the smart card as comparable with the personal computer – representing a milestone in the process of decentralising and personalising information services. Credito Valtellinese, which conducted the experiment, wanted to test the innovatory marketing potential of the card in addition to its technical aspects. The experiment involved 4,500 people and lasted 45 days. Over the period, cards were used on average 3.3 times. It is highly relevant that customers made considerable use of the ability to review the information on their cards.

Mastercard has carried out extensive research on consumer reaction to different aspects of smart cards, particularly the technology, the questions of fraud and PINs, the promotional benefits of the card, and the potential for new applications. Fraud emerged as being a concern for customers, in that 85% believed that they, rather than retailers or banks, paid for fraud. The research also showed that customers would use PINs, if they knew them, and there was a strong preference for self-selection of PINs.

The research suggested that smart cards had a better image than traditional cards, though this is likely to be a transient response to 'something new'. Consumers believed that the greatest advantage was increased security. 67% believed that the smart card was more secure than magnetic stripe cards. 60% of customers also felt that the smart card gave quicker payment authorisation. This was purely a perception, since in the experiment, conventional authorisation procedures were used. 42% of Mastercard customers believed that they would use the card more readily if additional features were added, with particular interest being expressed in full transaction records, and emergency medical and health information.

Indeed, the various experiments so far have given a base of information about consumer reaction to smart cards. Their opinions can be classified under reaction to debit rather than credit usage, reaction to information, reaction to convenience, reaction to expanded services, and reaction to technology. Convenience, security and information are the three most important attributes for general payment card customers. This is borne out by smart card research.

## 9.2  REACTION TO DEBIT RATHER THAN CREDIT

Research shows that, contrary to generally held opinion, Europeans are not against using credit, but are not keen to pay dearly for it. According to *People and Payments* [2] which surveyed over 6,000 consumers throughout Europe, "Europeans might well become more relaxed about using credit if the credit card product included built-in barriers against over-spending ... 74% of Europeans believe that in their current form, credit cards make it too easy to overspend." This presents an opportunity to smart card providers to build-in spending limitation to smart card users. But research also shows that there is a significant minority who are happier to use a debit card rather than a credit card.

## 9.3   REACTION TO CONVENIENCE

All experiments show that customers respond positively to increased convenience. This tells us nothing new. Laser cards for payphones are penetrating the market because of the advantages of carrying a card rather than heavy coins. Embossed credit cards are already accepted in many taxi services, to an enthusiastic response from business users. Convenience seekers are not very price sensitive. Convenience transcends cost. The possible uptake of the smart card for convenience services might follow the pattern for that other new high-tech product, cellular telephones. Indeed, the spectacular uptake of cellular telephones has demonstrated once again that if a new service offers greater convenience than previously was available, customers will pay heavily to get it. But the smart card has existing keen competition as far as increasing convenience is concerned. In financial services, customers have already shown a healthy response to being able to pay over the telephone simply by quoting their credit card number. The smart card will have to be considerably better if it is to take over the remote payment market.

In the United States [3] as in Europe, a high proportion of credit card holders state that they use the card for payment convenience rather than to get credit. Half of all card holders claim that they pay off the full outstanding balance on their card account each month. These customers have no inhibitions about instant debiting of their accounts, and are, indeed, far more interested in knowing their present financial position than in deferring payment.

## 9.4   REACTION TO INFORMATION

Research also shows that in Europe, consumers are keenly concerned about their personal wealth. 90% of all survey participants felt that financial information was a matter of great importance to them, and 63% considered themselves to be well-informed on financial matters. However, the smart card to date does not give the kind of information that the customer wants. " Even though the French consider themselves to have a technological lead, only 54% of them consider themselves to be well-informed."

Proportion of customers who consider
themselves well-informed on financial
matters

| | |
|---|---|
| Sweden | 93% |
| Netherlands | 92% |
| Spain | 37% |
| French | 54% |
| West Germany | 65% |
| United Kingdom | 70% |

Source: *People and Payments* [2]

A significant reason for interest in home banking, where such interest does exist, is the belief that it will give access to precise, up-to-date information.


## 9.5 REACTION TO SECURITY

Security has a high rating amongst consumers. Using cards rather than cash makes people feel more secure.

Card fraud is a concern for customers, and like the banks, they welcome ways of reducing the chances that they themselves will suffer loss due to card theft or copying. But the chip card is not likely to be adopted solely for its resistance to copying when other, cheaper, techniques have so far been ignored by the banking industry. Other services must be added. According to Frazer [4] "The smart card (in EFTPOS) is likely to stand or fall on its ability to store transaction data or support complex personal identification procedures. The extra cost of chip cards could be economically justified if they succeed in reducing disputed transactions. It is even possible that one day they could become mandatory by legislation if disputes become too serious a problem."

Signatures, unlike PINs, are not forgotten, and a written signature provides unquestionable proof of the transaction. Here the smart card can score, in its ability to store signature images.

The smart card can also improve the customer's position when transactions are disputed. The problem of disputed transactions is one of the most intractable and least considered of all those facing EFTPOS systems. Issuing transaction receipts proves nothing. If a customer denies that he made a transaction, he needs to hold documentary proof that he did *not* make that transaction. Absence of a receipt is not proof that the transaction did not take place. Cards which store sufficiently informative records in an unalterable way would go a long way to solving this problem −

provided that the customer himself is able to read the record, and he does not have to rely on the bank to read the card for him.


## 9.6    REACTION TO EXPANDED SERVICE

According to the report *People and Payments* [2], so far, European consumers have outpaced their bankers in adopting new attitudes towards their financial affairs. Variety does not confuse the man in the street. Consumers are well aware of the different kinds of payments services available to them, and have a firm opinion of how they can best use each of them. The European traveller in particular, is adapting fast to non-cash payment methods, and the business traveller fastest of all. The further they go away from home, the more they use cards. Consumers are aware that they must think ahead about how to pay when they are abroad. If they have ATM cards or credit cards they need to know whether there are facilities in the country they are going to for using the card. This explains the continuing popularity of travellers' cheques, which are very negotiable, yet insured against theft. But this does not mean a ready market exists for the smart card — in travel, what matters is that money is acceptable. People want proven methods.

Real markets exist for convenience in payment methods, and the assurance of payment in an emergency. Ease of replacement, acceptance in a foreign country, and usefulness for local purchases are all highly rated by consumers. Deferred payment — either with a 'grace period' before settlement, or with credit availability — both have relatively low ratings. There is hardly any demand at all for the status conferred by prestige payment products. Payment cards are not regarded as giving status.

But consumers must be made aware of real advantages. Early experiments showed that neither the public nor retailers wanted to use smart cards. To market them at all it was necessary to play down the technology in order to discover what new services consumers might be interested in. The following table indicates how customers want existing credit cards improved. ATM uptake has shown that customers respond positively to increased availability of services, both in time and in place.

Consumers will do business with any institution which can solve their financial problems. Most consumers are fairly happy with the banking services they get from their present bank. However, the young are less satisfied than the old, and the well-educated less satisfied than those who finished school early. Business travellers tend to be the most dissatisfied. What this means is that those who are most likely to generate ongoing profitability for the institutions serving them, are also the most unhappy with current services on offer. This means that the smart card may indeed offer a market entry mechanism for an institution wishing to break into the financial services marketplace.

| New services of interest from a credit card (2,233 respondents) | |
|---|---|
| proposed new service | proportion of card holders interested |
| increased protection | 52% |
| immediate information stored in card | 24% |
| quicker credit authorisation | 15% |
| none | 7% |
| (multiple answers permitted) | |

Source: Evans [5]

## 9.7   REACTION TO TECHNOLOGY

*People and Payments* [2] reports that bankers in Europe are more nervous than their customers about new methods of payment. "People enjoy the relaxation of rigid banking customs that comes with new technology. They are not unduly worried about the potential invasion of privacy and they are anxious to assert their right of choice." 45% of respondents feel comfortable using new banking products – only 33% feel uncomfortable. "Europeans welcome those advances in technology which will provide them with greater convenience and protection."

General research shows that market acceptance of a particular technology depends on market domination by that technology. Already, there are over 2 billion magnetic stripe cards in the marketplace, giving customers a service which is mostly considered satisfactory. Those countries where there is extensive magnetic stripe penetration are not likely to respond very positively to a smart card offering which merely duplicates or only slightly enhances what customers already have. The smart card, indeed, is more likely to find acceptance in countries where to date there is low card penetration. The situation can be compared with the development of the video recorder market in the United Kingdom.

Originally, the sales of video recorders were very sluggish. There were two competing systems, VHS and BETAMAX, each equally likely to be the ultimate standard. The major change in the marketplace came when the Open University ordered video recorders for each of its remote study locations. The choice was VHS. Thereafter, Open University students in vast numbers followed a now *de facto* standard, and after years of struggling to establish a hold in the marketplace, the BETAMAX system became obsolete. The smart card faces a similar uphill struggle in conventional marketplaces. Merely as an alternative technology, the smart card will not find acceptance.


## 9.8  SPECIAL MARKET SECTORS

In forecasting uptake of smart cards by consumers it is essential to be aware of the special features of today's marketplace that make it different from that of even twenty years ago.

The baby boom which took place in the twenty years following the second world war has contributed a major market sector now aged between midtwenties and midforties. During each step in their maturing, the baby boomers created a demand for new services and increased the success of many existing products and services. This group has specific characteristics that must be understood by any service supplier wishing to make a success in today's marketplace [6]. This group has a need for mass belonging. While they have wished to be different from their parents, they desire to be like their peers. Individually and collectively they have been characterised as self-conscious, assertive and idealistic. They are, though, very much a think-alike, do-alike generation. This means that if a new product can be accepted by this market, it will then penetrate quickly. They are a consuming group, wanting to live now, and not worrying about paying later. They are now in their peak consuming years.

This group is less price sensitive and more value sensitive than their parents. They are very convenience oriented.

A special group amongst baby boomers are the PC users. Many consumers hoped that home computers would help them manage their household financial affairs. As many as half of all consumers wanted to use PCs to simplify their personal finances, but when it came to putting it into practice, the general experience was that the results did not justify the effort. So consumers became disillusioned. This created a barrier to the successful introduction of computer-based financial planning. This desire for financial information and management represents a real opportunity for the promoters of the smart card.

This generation is not interested in traditional forms of banking. A 1984 American Banker survey found that 37% of respondents would be comfortable banking by mail, phone and ATM with a bank which no longer operated branches. There was a wide variation with age. Among 18–34 year olds, half would be comfortable banking without branches, but only 22% of those over 55 would be comfortable banking the high-tech way.

Income still has a significant effect on the frequency of card use. In general, the higher the income the more likely the cardholder is to use the card frequently. A high level of education is also a predictor of a high level of card use. In France, this is most marked, with those who have completed higher education four times as likely as those who have finished school early to use their cards more than ten times a month. Average use of credit cards by the entire group of credit cardholders surveyed in *People and Payments* was about four times per month. The table below sets out common payment methods. In addition, 2% of customers say they are likely to use a store card to buy a dress or jacket − though this low response may be due to confusion between store cards and credit cards, which are becoming more and more indistinguishable. It is also likely that the importance of petrol payment cards is underestimated in the survey. 13% of consumers will use extended shop credit to buy a refrigerator.

Likely payment methods used for everyday purchases (% of all consumers)

| | dress or jacket | restaurant meal | refrigerator | petrol | intercity rail ticket |
|---|---|---|---|---|---|
| cash | 62 | 75 | 33 | 59 | 66 |
| cheque | 27 | 14 | 33 | 14 | 18 |
| credit card | 5 | 5 | 5 | 6 | 3 |

Compiled from *People and Payments* [2]

But service providers must consider the environment as well as the people. In the United States there was, before the revision of tax practice, a strong demand for financial advisory services, particularly tax advice and assistance with tax returns.

## 9.9 THE FUTURE

So far, this chapter has addressed the question of what the consumer thinks will be the impact of the smart cards on financial services, relating this particularly to the market for payment cards. But the chapter title is 'the user's view'. It is worth focusing on that word 'user'. In fact, this is the key to success or failure in the marketplace, because people buy uses, not things. Even the unworldly scientist who is so often thought of as being interested in gadgets for their own sake is, in fact, 'buying' their use as a stimulus for the pleasure of intellectual fascination. But the average man in the street does not share this kind of excitement. Before *he* buys, he will ask the question 'What does it *do* for *me*?'

Outside banking, there is increasing use of smart cards as secure tokens for access control. Personal, portable processing, and information management are also becoming more commonplace. The predominant position of banking in the early applications of the smart card lies behind the generally accepted view that market segmentation and customer classification for smart card usage should mirror those used in general financial services marketing. But the well-established category descriptors that roll off the tongues of bank marketing personnel are not necessarily appropriate to the potential emerging markets for smart cards. Financial marketing segmentation depends on financial function (what people want to do with a service) – such as take a mortgage loan, get credit, do financial budgeting, make regular payments, buy at the supermarket, and so on. However, the smart card is not a function, but a form (the shape of the service). When used to make a payment, a smart card is to a cheque what an igloo is to a gypsy caravan. When used for medical records, it bears no relation to a cheque, but is very like a hospital file. When used to control access, it is nothing like a hospital file, but very like a key.

When we are defining the market for a new style of key, we classify potential customer groups by security needs – not by age, sex, marital status, profession and income. So, with the smart card, to define appropriate market segments to project future uptake and usage, we must have a clear idea of the functions that the card is providing for the consumer. It makes sense to think of different market classifications when we are dealing with payment cards, information users, financial planners, security seekers, and so on.

There are a number of common schemes for segmenting consumer markets. Geographic segmentation is a very traditional approach, and works best in markets which are dependent on social stratification. Geographic segmentation is based on the belief that similar people tend to group together. Demographics, such as age and income, are very commonly used in the financial services sector. This type of segmentation relies on the belief that customer needs and desires depend on professional status, and change with each generation. Geo-demographic segmentation combines these two approaches. Each of these three approaches is useful for historic analysis of customer behaviour patterns, but is weak when the service supplier wants to forecast likely uptake of new products and services, since they do not explain customer buying motivation.

Product usage and profitability are occasionally used for customer analysis, but only when the service supplier already has the ability to collect ongoing information on customer behaviour. When available, this approach can be very useful for predicting reaction to new products, since it is possible to use it to get insight into what motivates the customer. Needs and benefits analysis, when appropriate data can be gathered, assists the service provider in designing appropriate new products for his chosen marketplace. It is less useful when seeking a new market for an existing product. Psychographics, or lifestyle segmentation, presents the most relevant approach for marketing new services such as the smart card could offer, but requires highly creative input from promotional and advertising personnel to identify the relevant classifications appropriate to selling emerging technologies.

Forecasting the future markets for the new applications of smart cards means, first of all, clearly identifying what function the new application fulfills, secondly, identifying a comparable product already on the market (if such exists), thirdly, examining the market research on consumer attitudes to the existing product, and so identifying the particular advantage of the smart card to carry out that existing function. In this way, product designers can identify the unique selling point (USP) of the smart card in each new application. This USP is then the basis for the promotion of the smart card in the new marketplace.

The full potential of the smart card will only be realised when we have a clear understanding of the human needs that the gadget can fulfill. Normally, in marketing, the first thing is to define the market, understand its needs, and so develop the new products that will satisfy those needs. Marketing new technology is different. The inventions of science are by no means always mothered by necessity. Often they are mothered by serendipity — the happy coincidence of curiosity and creative imagination to produce something never before imagined. The negative way of looking at these happy chances is that they are 'solutions looking for a problem'. The positive way is to regard them as opportunities for new ways of doing things — or indeed, ways of doing new things. The personal stereo is a perfect example of a chance observation being developed into a major way of doing something new; the designer one day realised that his children carried their tape player about with them as they got on with other things, and used headphones so that they could each listen to his own choice of music, without interruption.

The real future of the smart card may well in retrospect be based on a similar inspiration. One thing is certain — merely as a gadget, as a scientific curiosity, the smart card has no commercial future. As such it would at best be like the Rubik's cube or cabbage patch dolls, here for a season and then forgotten.

This book identifies many categories of use — a token, a record store or filing mechanism, a processor, a communicating device, a prestige indicator, or a base for coordinating any combination of these functions.

None of these is unique to the smart card. In any individual use, success for the smart card will depend on clearly indicating that the smart card is better than existing alternatives, perhaps by being cheaper, or more reliable, or more flexible, or more personalised, or more versatile, or easier to use, or nicer to handle — or anything, indeed, that makes human beings feel that using the smart card makes life richer.

Life can be made richer either by making it less bad (reducing cost or pain) or by making it more good (increasing profit or pleasure). Individuals tend to react more strongly to new products that increase pleasure rather than reduce cost. People will be enthusiastic about Cordon Bleu frozen television meals, for example, while being passive about cost-saving soya mince pie filling. When the smart card gives something distinctly and uniquely positive, it is ready to take off as a commercial product. Perhaps then, financial institutions need not concern themselves with the cost of supplying the cards. Instead, customers might come to the banks requesting that the bank card functions be installed on the personal smart cards that they themselves have bought for their own uses.

## REFERENCES

[1] (1985) *Smart Cards – from France to the World* Lafferty Publications, London.
[2] (1987) *People and Payments* Lafferty Publications, London.
[3] Canner *et al.* (1985) 'Recent developments in credit card holding and use' *Journal of Retail Banking*. Fall 1985.
[4] Frazer (1985) 'EFTPOS – issues and insights' *Journal of Retail Banking*. Summer 1985.
[5] Evans (1987) 'A European study of consumer payments' *Journal of Retail Banking*. Fall 1987.
[6] Koehn (1986) 'The baby boom market: 2001' *Journal of Retail Banking*. Winter 86/87.

## BIBLIOGRAPHY

Esters and Christenson (1986) 'Market segmentation' *Journal of Retail Banking*. Winter 86/87.
Cox and Lasley (1984) 'Customer attitudes, behaviour and satisfaction' *Journal of Retail Banking*. Winter 1984.
Alyanakian (1985) 'Norwegian telebank experiment with smart card' *Proceedings of the EFTPOS & Home Banking Conference*, Edinburgh 1985.
Kehellen (1985) 'Credit card fraud control activities and smart card testing' *Proceedings of the EFTPOS & Home Banking Conference*, Edinburgh 1985.
Goslar (1986) 'Smart cards and payment systems strategies' *Proceedings of the EFTPOS & Home Services Conference*, Edinburgh 1986.
Malecki (1986) 'Pros and cons of smart vs. laser vs. mag. stripe cards' *Proceedings of the EFTPOS & Home Services Conference*, Edinburgh 1986.

# Index

Page numbers in italics refer to illustrations.